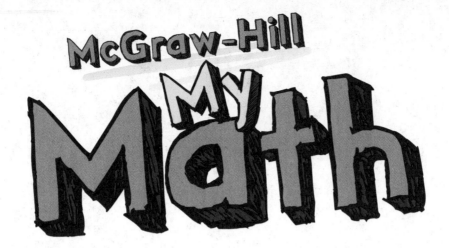

Welcome to *My Math* —your very own math book! You can write in it—in fact, you are encouraged to write, draw, circle, explain, and color as you explore the exciting world of mathematics. Let's get started. Grab a pencil and finish each sentence.

My name is _____.

My favorite color is _____.

My favorite hobby or sport is _____.

My favorite TV program or video game is
_____.

My favorite class is _____.

mhmymath.com

Copyright © 2018 McGraw-Hill Education

All rights reserved. No part of this publication may be reproduced or distributed in any form or by any means, or stored in a database or retrieval system, without the prior written consent of McGraw-Hill Education, including, but not limited to, network storage or transmission, or broadcast for distance learning.

STEM McGraw-Hill is committed to providing instructional materials in Science, Technology, Engineering, and Mathematics (STEM) that give all students a solid foundation, one that prepares them for college and careers in the 21st century.

Send all inquiries to:
McGraw-Hill Education
8787 Orion Place
Columbus, OH 43240

ISBN: 978-0-07-905761-7 (**Volume 1**)
MHID: 0-07-905761-6

Printed in the United States of America.

6 7 8 9 LWI 23 22 21 20

Understanding by Design® is a registered trademark of the Association for Supervision and Curriculum Development ("ASCD").

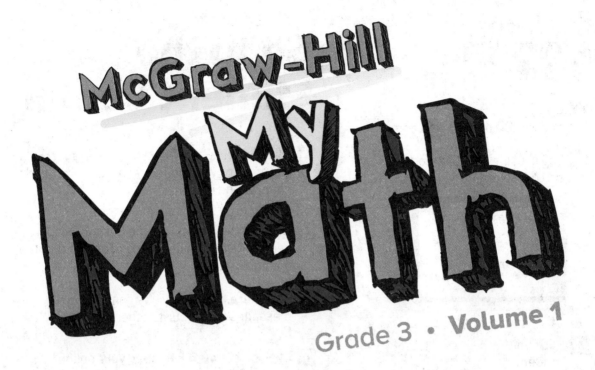

Grade 3 • Volume 1

Authors:
Carter • Cuevas • Day • Malloy
Altieri • Balka • Gonsalves • Grace • Krulik • Molix-Bailey
Moseley • Mowry • Myren • Price • Reynosa • Santa Cruz
Silbey • Vielhaber

GO digital ▶ connectED.mcgraw-hill.com

▶ Log In

1 Go to **connectED.mcgraw-hill.com**.

2 Log in using your username and password.

3 Click on the Student Edition icon to open the Student Center.

▶ Go to the Student Center

4 Click on Menu, then click on the **Resources** tab to see all of your online resources arranged by chapter and lesson.

5 Click on the **eToolkit** in the Lesson Resources section to open a library of eTools and virtual manipulatives.

6 Look here to find any assignments or messages from your teacher.

7 Click on the **eBook** to open your online Student Edition.

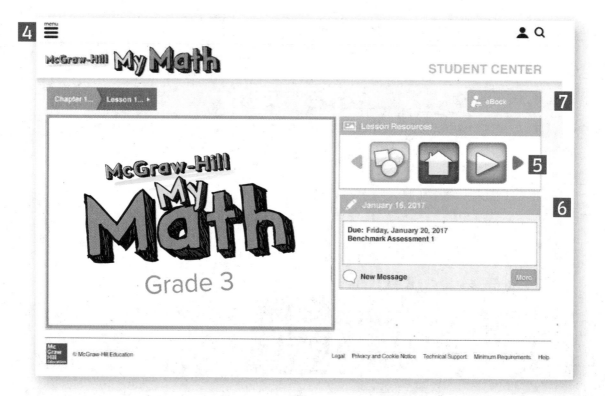

Username: _____

Password: _____

▶ Explore the eBook!

8 Click the **speaker icon** at the top of the eBook pages to hear the page read aloud to you.

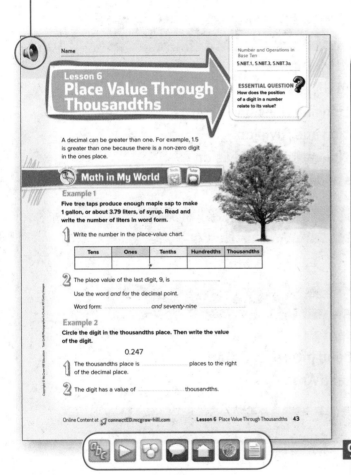

More resources can be found by clicking the icons at the bottom of the eBook pages.

 Practice and review your Vocabulary.

 Animations and videos allow you to explore mathematical topics.

 Explore concepts with eTools and virtual manipulatives.

 Personal Tutors are online virtual teachers that walk you through the steps of the lesson problems.

9 eHelp helps you complete your homework.

 Explore these fun digital activities to practice what you learned in the classroom.

 Worksheets are PDFs for Math at Home, Problem of the Day, and Fluency Practice.

Contents in Brief
Organized by Domain

Number and Operations in Base Ten

Chapter 1   Place Value
Chapter 2   Addition
Chapter 3   Subtraction

Operations and Algebraic Thinking

Chapter 4   Understand Multiplication
Chapter 5   Understand Division
Chapter 6   Multiplication and Division Patterns
Chapter 7   Multiplication and Division
Chapter 8   Apply Multiplication and Division
Chapter 9   Properties and Equations

Number and Operations — Fractions

Chapter 10   Fractions

Measurement and Data

Chapter 11   Measurement
Chapter 12   Represent and Interpret Data
Chapter 13   Perimeter and Area

Geometry

Chapter 14   Geometry

connectED.mcgraw-hill.com

Chapter 1: Place Value

Number and Operations in Base Ten

ESSENTIAL QUESTION
How can numbers be expressed, ordered, and compared?

Getting Started

My Chapter Project	2
Am I Ready?	3
My Math Words	4
My Vocabulary Cards	5
My Foldable **FOLDABLES**	7

Lessons and Homework

Lesson 1	Place Value Through Thousands	9
Lesson 2	Compare Numbers	15
Lesson 3	Order Numbers	21
Check My Progress		27
Lesson 4	Round to the Nearest Ten	29
Lesson 5	Round to the Nearest Hundred	35
Lesson 6	Problem-Solving Investigation: Use the Four-Step Plan	41

Wrap Up

My Chapter Review	47
Reflect	50
Performance Task	50PT1

Look for this!
Click online and you can watch videos that will help you learn the lessons.

connectED.mcgraw-hill.com

Number and Operations in Base Ten

ESSENTIAL QUESTION
How can place value help me add larger numbers?

Chapter 2: Addition

Getting Started

My Chapter Project	52
Am I Ready?	53
My Math Words	54
My Vocabulary Cards	55
My Foldable FOLDABLES	59

Lessons and Homework

Lesson 1	Addition Properties	61
Lesson 2	Patterns in the Addition Table	67
Lesson 3	Addition Patterns	73
Lesson 4	Add Mentally	79
Check My Progress		85
Lesson 5	Estimate Sums	87
Lesson 6	Hands On: Use Models to Add	93
Lesson 7	Add Three-Digit Numbers	99
Check My Progress		105
Lesson 8	Add Four-Digit Numbers	107
Lesson 9	Problem-Solving Investigation: Reasonable Answers	113

Wrap Up

Fluency Practice	119
My Chapter Review	121
Reflect	124
Performance Task	124PT1

connectED.mcgraw-hill.com

Chapter 3 Subtraction

Number and Operations in Base Ten

ESSENTIAL QUESTION
How are the operations of subtraction and addition related?

Getting Started

My Chapter Project	126
Am I Ready?	127
My Math Words	128
My Vocabulary Cards	129
My Foldable **FOLDABLES**	131

Lessons and Homework

Lesson 1	Subtract Mentally	133
Lesson 2	Estimate Differences	139
Lesson 3	Problem-Solving Investigation: Estimate or Exact Answer	145
	Check My Progress	151
Lesson 4	Hands On: Subtract with Regrouping	153
Lesson 5	Subtract Three-Digit Numbers	159
Lesson 6	Subtract Four-Digit Numbers	165
Lesson 7	Subtract Across Zeros	171

Wrap Up

Fluency Practice	177
My Chapter Review	179
Reflect	182
Performance Task	182PT1

Look for this! Click online and you can get more help while doing your homework.

connectED.mcgraw-hill.com

Operations and Algebraic Thinking

ESSENTIAL QUESTION
What does multiplication mean?

Chapter 4
Understand Multiplication

Getting Started

My Chapter Project **184**
Am I Ready? **185**
My Math Words **186**
My Vocabulary Cards **187**
My Foldable FOLDABLES **191**

Lessons and Homework

Lesson 1 Hands On: Model Multiplication **193**
Lesson 2 Multiplication as Repeated Addition **199**
Lesson 3 Hands On: Multiply with Arrays **205**
Lesson 4 Arrays and Multiplication **211**
Check My Progress **217**
Lesson 5 Problem-Solving Investigation:
　　　　　　Make a Table **219**
Lesson 6 Use Multiplication to
　　　　　　Find Combinations **225**

Wrap Up

My Chapter Review **231**
Reflect **234**
Performance Task **234PT1**

connectED.mcgraw-hill.com

Chapter 5: Understand Division

Operations and Algebraic Thinking

ESSENTIAL QUESTION
What does division mean?

Getting Started

My Chapter Project	236
Am I Ready?	237
My Math Words	238
My Vocabulary Cards	239
My Foldable **FOLDABLES**	243

Lessons and Homework

Lesson 1	Hands On: Model Division	245
Lesson 2	Division as Equal Sharing	251
Lesson 3	Relate Division and Subtraction	257
Check My Progress		263
Lesson 4	Hands On: Relate Division and Multiplication	265
Lesson 5	Inverse Operations	271
Lesson 6	Problem-Solving Investigation: Use Models	277

Wrap Up

My Chapter Review	283
Reflect	286
Performance Task	286PT1

Look for this! Click online and you can find tools that will help you explore concepts.

connectED.mcgraw-hill.com

Operations and Algebraic Thinking

ESSENTIAL QUESTION
What is the importance of patterns in learning multiplication and division?

Chapter 6: Multiplication and Division Patterns

Getting Started

My Chapter Project	288
Am I Ready?	289
My Math Words	290
My Vocabulary Cards	291
My Foldable **FOLDABLES**	293

Lessons and Homework

Lesson 1	Patterns in the Multiplication Table	295
Lesson 2	Multiply by 2	301
Lesson 3	Divide by 2	307
Lesson 4	Multiply by 5	313
Lesson 5	Divide by 5	319
Check My Progress		325
Lesson 6	Problem-Solving Investigation: Look for a Pattern	327
Lesson 7	Multiply by 10	333
Lesson 8	Multiples of 10	339
Lesson 9	Divide by 10	345

Wrap Up

Fluency Practice	351
My Chapter Review	353
Reflect	356
Performance Task	356PT1

connectED.mcgraw-hill.com

Chapter 7 Multiplication and Division

Operations and Algebraic Thinking

ESSENTIAL QUESTION
What strategies can be used to learn multiplication and division facts?

Getting Started

My Chapter Project	358
Am I Ready?	359
My Math Words	360
My Vocabulary Cards	361
My Foldable **FOLDABLES**	363

Lessons and Homework

Lesson 1	Multiply by 3	365
Lesson 2	Divide by 3	371
Lesson 3	Hands On: Double a Known Fact	377
Lesson 4	Multiply by 4	383
Lesson 5	Divide by 4	389
	Check My Progress	395
Lesson 6	Problem-Solving Investigation: Extra or Missing Information	397
Lesson 7	Multiply by 0 and 1	403
Lesson 8	Divide with 0 and 1	409

Wrap Up

Fluency Practice	415
My Chapter Review	417
Reflect	420
Performance Task	420PT1

Look for this! Click online and you can watch a teacher solving problems.

connectED.mcgraw-hill.com

Operations and Algebraic Thinking

ESSENTIAL QUESTION
How can multiplication and division facts with smaller numbers be applied to larger numbers?

Chapter 8
Apply Multiplication and Division

Getting Started

My Chapter Project	422
Am I Ready?	423
My Math Words	424
My Vocabulary Cards	425
My Foldable **FOLDABLES**	427

Lessons and Homework

Lesson 1	Multiply by 6	429
Lesson 2	Multiply by 7	435
Lesson 3	Divide by 6 and 7	441
Check My Progress		447
Lesson 4	Multiply by 8	449
Lesson 5	Multiply by 9	456
Lesson 6	Divide by 8 and 9	461
Check My Progress		467
Lesson 7	Problem-Solving Investigation: Make an Organized List	469
Lesson 8	Multiply by 11 and 12	475
Lesson 9	Divide by 11 and 12	481

Wrap Up

Fluency Practice	487
My Chapter Review	489
Reflect	492
Performance Task	492PT1

connectED.mcgraw-hill.com

Chapter 9 Properties and Equations

Operations and Algebraic Thinking

ESSENTIAL QUESTION
How are properties and equations used to group numbers?

Getting Started

My Chapter Project	494
Am I Ready?	495
My Math Words	496
My Vocabulary Cards	497
My Foldable **FOLDABLES**	499

Lessons and Homework

Lesson 1	Hands On: Take Apart to Multiply	501
Lesson 2	The Distributive Property	507
Lesson 3	Hands On: Multiply Three Factors	513
Lesson 4	The Associative Property	519
Check My Progress		525
Lesson 5	Write Expressions	527
Lesson 6	Evaluate Expressions	533
Lesson 7	Write Equations	539
Lesson 8	Solve Two-Step Word Problems	545
Lesson 9	Problem-Solving Investigation: Use Logical Reasoning	551

Wrap Up

My Chapter Review	557
Reflect	560
Performance Task	560PT1

Look for this!
Click online and you can find activities to help build your vocabulary.

connectED.mcgraw-hill.com

Number and Operations — Fractions

ESSENTIAL QUESTION
How can fractions be used to represent numbers and their parts?

Chapter 10 Fractions

Getting Started

My Chapter Project 562
Am I Ready? 563
My Math Words 564
My Vocabulary Cards 565
My Foldable **FOLDABLES** 567

Lessons and Homework

Lesson 1 Unit Fractions 569
Lesson 2 Part of a Whole 575
Lesson 3 Part of a Set 581
Lesson 4 Problem-Solving Investigation: Draw a Diagram 587
Check My Progress 593
Lesson 5 Hands On: Fractions on a Number Line 595
Lesson 6 Equivalent Fractions 601
Lesson 7 Fractions as One Whole 607
Lesson 8 Compare Fractions 613

Wrap Up

My Chapter Review 619
Reflect 622
Performance Task 622PT1

connectED.mcgraw-hill.com

Chapter 11 Measurement

Measurement and Data

ESSENTIAL QUESTION
Why do we measure?

Getting Started

My Chapter Project 624
Am I Ready? 625
My Math Words 626
My Vocabulary Cards 627
My Foldable **FOLDABLES** 631

Lessons and Homework

Lesson 1 Hands On: Estimate and Measure Capacity 633
Lesson 2 Solve Capacity Problems 639
Lesson 3 Hands On: Estimate and Measure Mass ... 645
Lesson 4 Solve Mass Problems 651
Check My Progress 657
Lesson 5 Tell Time to the Minute 659
Lesson 6 Time Intervals 665
Lesson 7 Problem-Solving Investigation: Work Backward 671

Wrap Up

My Chapter Review 677
Reflect 680
Performance Task 680PT1

connectED.mcgraw-hill.com

Measurement and Data

ESSENTIAL QUESTION
How do we obtain useful information from a set of data?

Chapter 12 Represent and Interpret Data

Getting Started

My Chapter Project 682
Am I Ready? 683
My Math Words 684
My Vocabulary Cards 685
My Foldable **FOLDABLES** 689

Lessons and Homework

Lesson 1 Collect and Record Data 691
Lesson 2 Draw Scaled Picture Graphs 697
Lesson 3 Draw Scaled Bar Graphs 703
Lesson 4 Relate Bar Graphs to
Scaled Picture Graphs 709
Lesson 5 Draw and Analyze Line Plots 715
Check My Progress 721
Lesson 6 Hands On: Measure to Halves
and Fourths of an Inch 723
Lesson 7 Collect and Display Measurement Data ... 729
Lesson 8 Problem-Solving Investigation:
Solve a Simpler Problem 735

Wrap Up

My Chapter Review 741
Reflect 744
Performance Task 744PT1

connectED.mcgraw-hill.com

Chapter 13 Perimeter and Area

Measurement and Data

ESSENTIAL QUESTION
How are perimeter and area related and how are they different?

Getting Started

My Chapter Project	746
Am I Ready?	747
My Math Words	748
My Vocabulary Cards	749
My Foldable FOLDABLES	751

Lessons and Homework

Lesson 1	Hands On: Find Perimeter	753
Lesson 2	Perimeter	759
Lesson 3	Hands On: Understand Area	765
Lesson 4	Measure Area	771
Check My Progress		777
Lesson 5	Hands On: Tile Rectangles to Find Area	779
Lesson 6	Area of Rectangles	785
Lesson 7	Hands On: Area and the Distributive Property	791
Lesson 8	Area of Composite Figures	797
Check My Progress		803
Lesson 9	Area and Perimeter	805
Lesson 10	Problem-Solving Investigation: Draw a Diagram	811

Wrap Up

My Chapter Review	817
Reflect	820
Performance Task	820PT1

connectED.mcgraw-hill.com

Geometry

ESSENTIAL QUESTION
How can geometric shapes help me solve real-world problems?

Chapter 14 Geometry

Getting Started

My Chapter Project	822
Am I Ready?	823
My Math Words	824
My Vocabulary Cards	825
My Foldable FOLDABLES	831

Lessons and Homework

Lesson 1	Hands On: Angles	833
Lesson 2	Polygons	839
Lesson 3	Hands On: Triangles	845
Lesson 4	Quadrilaterals	851
Check My Progress		857
Lesson 5	Shared Attributes of Quadrilaterals	859
Lesson 6	Problem-Solving Investigation: Guess, Check, and Revise	865
Lesson 7	Partition Shapes	871

Wrap Up

My Chapter Review	877
Reflect	880
Performance Task	880PT1

xx

Chapter 1 Place Value

ESSENTIAL QUESTION

How can numbers be expressed, ordered, and compared?

Let's Travel!

Watch a video!

Name _____

MY Chapter Project

Book Count

Day 1

1. Book subject category: _____

2. Work together to find the number of books in the library that belong to your book subject category.

3. Complete the place-value chart below with the number of books in your book subject category.

thousands	hundreds	tens	ones
,			

4. Think of a comparison question you could ask your classmates. Sample questions: Which category has more books, fiction or science? Is the number of books in the mathematics category more or less than the number of books in the biographies category? How much more/less?

Question: _____

_____

Day 2

1. Fill in the class place-value chart with the number of books in your subject category.

2. Ask the class your comparison question and answer other groups' comparison questions.

Name _____

Write each number.

1. | hundreds | tens | ones |
   |----------|------|------|
   | | 1 | 4 |

   _____

2. | hundreds | tens | ones |
   |----------|------|------|
   | | 3 | 3 |

   _____

3. | hundreds | tens | ones |
   |----------|------|------|
   | 1 | 1 | 0 |

   _____

4. 1 ten 5 ones _____   5. 1 hundred 2 ones _____

Write the number of tens and ones in each number.

6. 12 _____   7. 26 _____

Compare. Use >, <, or =.

8. 70 ◯ 61   9. 98 ◯ 99   10. 155 ◯ 55

11. What number is 10 fewer than 66?   12. What number is 100 greater than 800?

13. Deidra has three cards each with a value of 10 and two cards each with a value of 1. Raul has three cards each with a value of 1 and two cards each with a value of 10. Whose cards have the lesser value? Explain.

_____

_____

Shade the boxes to show the problems you answered correctly.

How Did I Do? | 1 | 2 | 3 | 4 | 5 | 6 | 7 | 8 | 9 | 10 | 11 | 12 | 13 |

Online Content at connectED.mcgraw-hill.com

Name _____

MY Math Words   Vocab

Review Vocabulary

hundreds   is equal to (=)   is greater than (>)

is less than (<)   ones   tens

Making Connections
Use the review vocabulary to complete the graphic organizer. You will not use every word. You will use a symbol in one of your answers.

5 is in the _____ place.

4 is in the _____ place.

546

546 ◯ 546

6 is in the _____ place.

Write a sentence using one or more of the review vocabulary words.

_____

4 Chapter 1 Place Value

MY Vocabulary Cards

Lesson 1–1

digit

0 1 2 3 4 5 6 7 8 9

Lesson 1–1

expanded form

672 = 600 + 70 + 2

Lesson 1–1

place value

thousands	hundreds	tens	ones
4	5	2	9

4,000 **500** **20** **9**

Lesson 1–4

round

36 is closer to 40

Lesson 1–1

standard form

3,491

Lesson 1–1

word form

six thousand,
four hundred ninety-nine

Ideas for Use

- During this school year, create a separate stack of cards for key math verbs such as *round*. Understanding these verbs will help you in your problem solving.

- Practice your penmanship! Write each word in cursive.

A way of writing a number as a sum that shows the value of each digit.

Explain what *expanded* means in this sentence: *The balloon expanded as it filled with air.*

Any symbol used to write whole numbers.

Write a three-digit number.

To change the value of a number to one that is easier to work with.

Round is a word with multiple meanings. Choose another meaning of *round*, and use it in a sentence.

The value given to a *digit* by its place in a number.

Write a number in which 6 is in the tens place and in the hundreds place.

The form of a number that uses written words.

Write the number 4,274 in word form.

The usual way of writing a number that shows only its digits, no words.

What is another way to write a number?

MY Foldable

FOLDABLES Follow the steps on the back to make your Foldable.

Round to 10s   Round to 100s

563
write
round

115
write
round

6,449
write
round

8,076
write
round

FOLDABLES® Study Organizer

Rounding Rules

1. Circle the digit to be rounded.
2. Look at the digit to the right of the place being rounded.
3. If the digit is less than 5, do not change the circled digit. If the digit is 5 or greater, add 1 to the circled digit.
4. Replace all digits after the circled digit with zeros.

USE PLACE VALUE TO ROUND

Name _____

Lesson 1
Place Value Through Thousands

ESSENTIAL QUESTION
How can numbers be expressed, ordered, and compared?

A **digit** is any symbol used to write whole numbers. The numbers 0, 1, 2, 3, 4, 5, 6, 7, 8, and 9 are all digits. The **place value** of a digit tells what value it has in a number.

Math in My World

Example 1

The height of the Statue of Liberty from the top of the base to the top of the torch is **1,813** inches. Identify the place of the highlighted digit in 1,813. Then write the value of the digit.

You can use 10 hundreds to show 1,000.

10 hundreds → 1 thousand

Model 1,813 with base-ten blocks.

Record each digit in the place value chart.

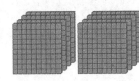

1 thousand   8 hundreds   1 ten   3 ones

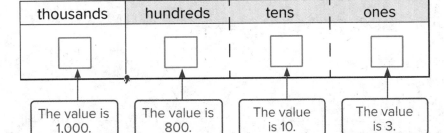

thousands	hundreds	tens	ones

The value is 1,000.   The value is 800.   The value is 10.   The value is 3.

The highlighted digit, 1, is in the _____ place. Its value is _____.

Online Content at connectED.mcgraw-hill.com

Example 2

If ten people climb the stairs to the top of the Statue of Liberty and back down, they will have walked 7,080 steps. What are the values of the zeros in the number 7,080?

thousands	hundreds	tens	ones
7	0	8	0

A comma is placed between the thousands and hundreds place.

When 0 is used in a number, it is a place holder. In 7,080, there are two place holders, the zero in the _____ place and the zero in the _____ place. Their values are 0.

Numbers can be written in different ways. **Standard form** shows only the digits. **Expanded form** shows the sum of the value of the digits. **Word form** uses words.

Example 3

The distance from Mobile, Alabama to the Statue of Liberty is 1,215 miles. Write 1,215 three ways.

Standard Form: ☐ , ☐ ☐ ☐

Expanded Form: 1,000 + _____ + 10 + _____

Word Form: _____ thousand, _____ hundred _____

Guided Practice

1. Write 7,009 in word form.

   _____ thousand, _____

2. Write 856 in expanded form.

   _____ + _____ + _____

How do I tell the value of each digit in a number?

10 Chapter 1 Place Value

Name _____

Independent Practice

Write the highlighted digit's place and value.

3. 5**0**1 _____   4. 5,**7**72 _____   5. 1,0**2**0 _____

6. **4**,810 _____   7. 3,1**7**6 _____   8. 80**4** _____

Write each number in standard form.

9. 4,000 + 600 + 70 + 8 _____   10. 3,000 + 20 + 1 _____

11. seven thousand, six hundred forty-one _____

12. eight thousand, seven hundred sixty _____

Write each number in expanded form and word form.

13. 4,332 _____ + _____ + _____ + _____

14. 6,503 _____ + _____ + _____ + _____

15. A motorcycle costs $3,124. What is the value of each digit?

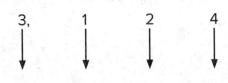

16. Write all of the three-digit numbers that have a 3 in the tens place and a 5 in the ones place.

Lesson 1 Place Value Through Thousands 11

Problem Solving

17. The model represents the number of days the South Pole does not have sunshine each year.

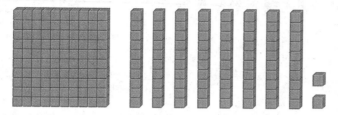

Circle the standard form of this number.

1,802   281   1,082   182

18. Martin earned 7,283 points while playing a video game. Circle the correct word form of this number.

seven thousand, eight hundred twenty-three

seven thousand, two hundred eighty-three

seven thousand, two hundred thirty-eight

Brain Builders

19. Processes &Practices 2 Use Number Sense Write the standard form for two thousand, thirteen and two hundred thirteen. How are they the same?

20. Building on the Essential Question Why does the position of each digit of a number matter? Explain your reasoning to a friend.

Name _____

MY Homework

Lesson 1

Place Value Through Thousands

Homework Helper

Need help? connectED.mcgraw-hill.com

There are 1,576 steps to the top of the Empire State Building. Look at the model. Write the number in standard form, expanded form, and word form.

 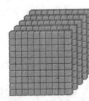

1 thousand   5 hundreds   7 tens   6 ones

Standard form: 1,576

Expanded form: 1,000 + 500 + 70 + 6

Word form: *one thousand, five hundred seventy-six*

Write the place and the value of the highlighted digit in 1,576.

A place-value chart helps to identify the place and value of a digit.

thousands	hundreds	tens	ones
1	⑤	7	6

The circled digit, 5, is in the hundreds place. Its value is 500.

Practice

Write each number in expanded form and word form.

1. 2,368   2,000 + _____ + 60 + _____

   two _____, three _____ sixty-eight

2. 6,204   _____ + _____ + _____

   _____ thousand, _____ hundred four

Lesson 1 My Homework   13

Write the highlighted digit's place and value.

3. 567 _____

4. 6,327 _____

5. 9,325 _____

6. 8,281 _____

Write each number in standard form.

7. 5,000 + 500 + 3 _____

8. 2,000 + 300 + 20 + 9 _____

9. 4,000 + 600 + 8 _____

10. 9,000 + 300 + 70 + 2 _____

Brain Builders

11. **Processes &Practices** Use Number Sense Kyle is in seat number 1,024. Sierra's seat number has the same number of thousands and tens as Kyle's seat number, but 2 more hundreds and 3 fewer ones than Kyle's seat number. What is Sierra's seat number? Describe how you found the seat number.

_____

_____

Vocabulary Check

Draw a line to connect each vocabulary word(s) to its definition.

12. word form • the value given to a digit by its place in a number

13. digit • the form of a number that uses written words

14. expanded form • the form of a number that shows the sum of the value of each digit

15. standard form • a symbol used to write a number

16. place value • the form of writing a number that shows only its digits

17. **Test Practice** Which represents the number that is ten more than *two thousand, seventy*?

Ⓐ 280   Ⓑ 2,008   Ⓒ 2,080   Ⓓ 2,800

Name _____

Lesson 2
Compare Numbers

ESSENTIAL QUESTION
How can numbers be expressed, ordered, and compared?

Math in My World

Symbol	Meaning
<	is less than
>	is greater than
=	is equal to

Example 1

The Tyee family is planning a road trip to the Grand Canyon. One route is 840 miles. A second route is 835 miles. Which route is shorter?

Compare 835 and 840.

One Way Use a place-value chart.

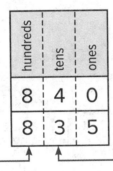

Both numbers have 8 hundreds.

840 has 4 tens, 835 has 3 tens, 4 tens > 3 tens

Another Way Use a number line.

830 831 832 833 834 835 836 837 838 839 840

less than (<) greater than (>)

835 **is to the left** of 840   840 **is to the right** of 835

835 ◯ 840   840 ◯ 835

Since _____ is less than _____, the _____ route is shorter.

Online Content at connectED.mcgraw-hill.com   Lesson 2 Compare Numbers 15

Example 2

During his hockey career, Mark Messier scored 1,887 points. Gordie Howe scored 1,850 points during his career. Which player scored a greater number of points during his career?

Compare 1,887 and 1,850.

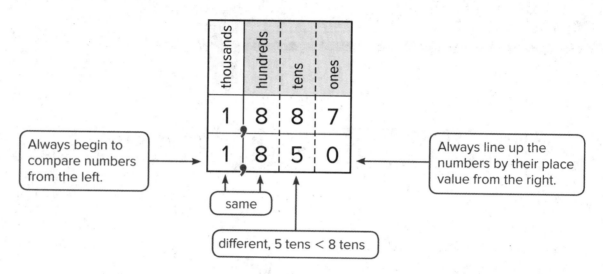

Always begin to compare numbers from the left.

Always line up the numbers by their place value from the right.

same

different, 5 tens < 8 tens

Since 8 > _____, 1,887 > _____.
Mark Messier scored the greater number of points.

Guided Practice

Which number is less? Complete the statement.

1.
hundreds	tens	ones
8	7	0
4	0	0

_____ < _____

2.
thousands	hundreds	tens	ones
9	6	3	0
6	4	0	3

_____ < _____

Talk MATH

Why is it not necessary to compare the ones digits in the numbers 365 and 378?

3. Use a number line to compare. Write >, <, or =.

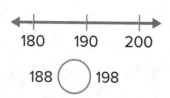

188 ◯ 198

16 **Chapter 1** Place Value

Name ....

Independent Practice

Compare. Use >, <, or =.

4. 604 ◯ 592 5. 188 ◯ 198 6. 1,000 ◯ 850

7. 999 ◯ 999 8. 1,121 ◯ 1,112 9. 6,573 ◯ 7,650

10. 2,644 ◯ 2,464 11. 1,000 ◯ 1,000 12. 3,039 ◯ 3,019

Circle the number that is greater. Then complete the statement.

13. 555   725   14. 800   700   15. 998   989

    _____ > _____   _____ > _____   _____ > _____

16. 931   8,310   17. 8,008 8,080   18. 2,753 2,735

    _____ > _____   _____ > _____   _____ > _____

Circle the number that is less. Then complete the statement.

19. 2,456 1,456   20. 3,052 3,050   21. 6,358 6,759

    _____ < _____   _____ < _____   _____ < _____

22. 5,317 5,318   23. 2,099 1,099   24. 1,321 1,231

    _____ < _____   _____ < _____   _____ < _____

Choose a number of your own to complete the statement.

25. 6,993 < _____   26. 2,209 = _____   27. _____ > 7,203

28. Circle all of the numbers that are greater than 4,109.

    5,109   4,019   4,191   4,091   4,108   4,110

Lesson 2 Compare Numbers 17

Problem Solving

29. The table shows the number of tickets sold for a movie. Which showing sold more tickets?

Revenge of Dinosaurs	
Showing	Tickets Sold
5:00 p.m.	235
7:00 p.m.	253

30. Circle the number that is less than 4,259.

4,260   4,300   4,209

Brain Builders

31. Processes &Practices 6 Explain to a Friend There are 165 students in the 3rd grade. There are 35 students in each of the three classes in the 2nd grade. Which grade has more students? Explain.

32. Processes &Practices 2 Use Number Sense Write the greatest and least 4-digit number you can make using each of the numerals 6, 3, 9, and 7 one time. Explain.

greatest _____   least _____

33. Building on the Essential Question How can I show how one number compares to another number? Explain to a friend.

18 Chapter 1 Place Value

Name _____

Lesson 2
Compare Numbers

Homework Helper

Need help? connectED.mcgraw-hill.com

The John Hancock Center, in Chicago, is 1,127 feet in height. The Aon Center is 1,136 feet in height. Which building is taller?

A place-value chart can help to compare numbers.

So, 1,127 is less than 1,136.

1,127 < 1,136

The Aon Center is taller.

thousands	hundreds	tens	ones
1	1	2	7
1	1	3	6

John Hancock Center → 1,127
Aon Center → 1,136

same

different, 3 tens > 2 tens

Use these symbols to compare.
< means **is less than**
> means **is greater than**
= means **is equal to**

Practice

Which number is less? Complete the statement.

1.
hundreds	tens	ones
6	9	6
6	8	7

_____ < _____

2.
thousands	hundreds	tens	ones
2	1	8	3
3	1	5	4

_____ < _____

Lesson 2 My Homework 19

Compare. Use >, <, or =.

3. 751 ◯ 715
4. 435 ◯ 543
5. 808 ◯ 880
6. 3,332 ◯ 3,332
7. 6,673 ◯ 6,376
8. 9,918 ◯ 9,819

Circle the number that is greater. Then complete the statement.

9. 3,322   3,332

_____ > _____

10. 1,877   1,788

_____ > _____

11. 2,727   2,772

_____ > _____

Circle the number that is less. Then complete the statement.

12. 5,642   5,426

_____ < _____

13. 4,017   4,071

_____ < _____

14. 6,310   6,231

_____ < _____

Brain Builders

15. **Processes & Practices 2** **Use Symbols** Last year there were 191 sunny days and 174 cloudy days. Were there more sunny or cloudy days last year? Compare using <, >, or =. Explain how you compared the numbers.

_____

_____

16. **Test Practice** Kara predicts the number of fish at an aquarium to be 1,776. Jake predicts that there are 1,715. The actual number is greater than the sum of their predictions. Which could be the actual number of fish at the aquarium?

Ⓐ 3,419
Ⓒ 3,491
Ⓑ 3,490
Ⓓ 3,499

20 **Need more practice?** Download Extra Practice at connectED.mcgraw-hill.com

Name ....

Lesson 3
Order Numbers

ESSENTIAL QUESTION
How can numbers be expressed, ordered, and compared?

Comparing numbers helps you to order them.

Math in My World

Example 1

The O'Dell family went on a whale-watching trip. They learned about different whales. The table shows the lengths of three whales. Order the lengths from *least* to *greatest*.

Average Lengths of Whales

Whale	Length (inches)
Orca Whale	264
Blue Whale	1,128
Humpback Whale	744

One Way Use a place-value chart.

Line up the numbers by their place value from the right. Compare from the left.

thousands	hundreds	tens	ones
	2	6	4
1	1	2	8
	7	4	4

1 thousand is the greatest number.

7 hundreds > 2 hundreds

Another Way Use a number line.

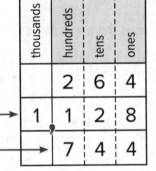

_____ < _____ < _____

The order from *least* to *greatest* is _____ inches, _____ inches, and _____ inches.

Online Content at connectED.mcgraw-hill.com

Lesson 3 Order Numbers 21

Example 2

The table shows the distances whales travel to feed in the summertime. This is called migration. Order these distances from *greatest* to *least*.

Whale Migration	
Whale	Distance (miles)
Humpback Whale	3,500
Gray Whale	6,200
Orca Whale	900

Use a place-value chart to line up the numbers by their place value. Compare from the left.

1 Compare the numbers with the greatest place value.

thousands	hundreds	tens	ones
3,	5	0	0
6,	2	0	0
	9	0	0

6 thousands > 3 thousands

2 Compare the remaining two numbers.

3 thousands > 0 thousands

The greatest number is _____. The second greatest number is _____.

_____ > _____ > _____

So, the whales' migration distances from greatest to least

are _____ miles, _____ miles, and _____ miles.

Guided Practice

Order the numbers from *least* to *greatest*.

1.
hundreds	tens	ones
3	9	
	6	8
3	3	2

2.
thousands	hundreds	tens	ones
	2	0	2
2,	2	0	2
	2	2	0

Talk MATH

Look at Exercise 2. Explain how you can tell which number is the greatest.

_____ ; _____ ; _____   _____ ; _____ ; _____

22 Chapter 1 Place Value

Name _____

Independent Practice

Order the numbers from *greatest* to *least*.

3. 303; 30; 3,003

4. 4,404; 4,044; 4,040

5. 1,234; 998; 2,134

6. 2,673; 2,787; 2,900

7. Animal Weights

1,000 pounds
530 pounds
345 pounds

8. Cars for Sale

$7,851
$6,342
$7,585

Order the numbers from *least* to *greatest*.

9. 60; 600; 6,006

10. 349; 343; 560

11. 3,587; 875; 2,435

12. 999; 1,342; 2,000

13. Pets for Sale

$395
$1,095
$1,090

14. Number of Students

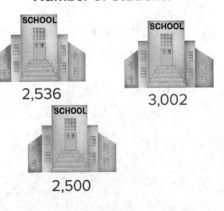

2,536
3,002
2,500

Lesson 3 Order Numbers 23

Problem Solving

15. During the season, Team A won 19 games, Team B won 40 games and Team C won 22 games. What place did each team earn for the season?

1st _____ 2nd _____ 3rd _____

16. Processes &Practices 2 Use Number Sense Write four numbers that could be ordered between these numbers.

59, _____, _____, _____, _____, 1,000

Brain Builders

17. Processes &Practices 8 Look for a Pattern Order the lengths of the rivers from longest to shortest. Explain your process.

River	Length (miles)
Arkansas	1,469
Mississippi	2,340
Missouri	2,540
Ohio	1,310
Red	1,290

18. Building on the Essential Question When and why does order matter? Explain your answer to a friend.

24 Chapter 1 Place Value

Name _____

MY Homework

Lesson 3
Order Numbers

Homework Helper

Need help? connectED.mcgraw-hill.com

The table shows the number of each type of car sold. Order the cars sold from *least* to *greatest*.

Type of Car	Cars Sold
Sports Utility	1,309
Sedan	1,803
Compact	1,117

Use a place-value chart.

Line up the numbers from the right.

Begin to compare from the left.

Number lines are another way to order numbers.

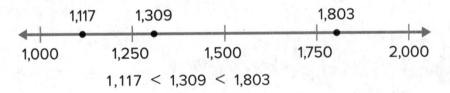

1,117 < 1,309 < 1,803

The order from *least* to *greatest* is compact, sports utility, and sedan.

Practice

Order the numbers from *least* to *greatest*.

1. 210; 182; 153

2. 1,692; 1,687; 1,685

3. 9,544; 9,455; 9,564

4. 653; 535; 335

Lesson 3 My Homework 25

Order the numbers from *greatest* to *least*.

5. **Electronics for Sale**

6. **Miles Traveled**

3,764 miles   3,647 miles   2,473 miles

Problem Solving

7. **Processes &Practices** ➊ **Make Sense of Problems** The Jacksons, Chens, and Simms each went on vacation. The Jacksons drove 835 miles. The Simms drove 947 miles and the Chens drove 100 miles further than the Jacksons. Which family drove the furthest on vacation?

Brain Builders

8. The addresses on Plum Street are out of order. Michael puts them in this order:

    7867, 8112, 7831

    Did he put them in order from least to greatest? Explain how you know.

9. **Test Practice** Four cars are parked in order from the least to greatest price. Which set of numbers shows the prices of the cars as they are parked?

    Ⓐ $7,659; $7,668; $8,985; $9,887   Ⓒ $8,985; $9,887; $7,668; $7,659

    Ⓑ $9,887; $8,985; $7,668; $7,659   Ⓓ $9,887; $8,985; $7,659; $7,668

Check My Progress

Vocabulary Check

1. Write each form of writing numbers in the boxes. Then, write two examples of numbers for each form.

   expanded form   standard form   word form

   Forms of Writing Numbers

Concept Check

Write the highlighted digit's place and value.

2. 729

3. 4,301

4. 6,291

Write each number in standard form.

5. six thousand, four hundred two

6. 2,000 + 90 + 3

Write each number in expanded form and word form.

7. 7,362

8. 3,035

Compare. Write >, <, or =.

9. 1,405 ◯ 1,450

10. 2,338 ◯ 2,338

11. 3,239 ◯ 2,993

Problem Solving

12. Last year, Gwen read 2,395 pages. Jorge read 3,093 pages. Stefanie read 2,935 pages. Write the numbers of pages read in order from *greatest* to *least*.

Brain Builders

13. Pierre writes that his family traveled 1,492 miles during summer vacation. He erases the 4 and writes a 7 in its place. What is the change in the value of this digit?

14. **Test Practice** Karly earned 2,489 credits on one flight. She earned 1,220 credits on another flight. What is the expanded form of the total number of credits she earned?

    Ⓐ 37 + 9

    Ⓑ 3,000 + 700 + 9

    Ⓒ 3,000 + 700 + 90

    Ⓓ 3,000 + 700 + 900 + 9

28 **Chapter 1** Place Value

Name ....

Lesson 4
Round to the Nearest Ten

ESSENTIAL QUESTION
How can numbers be expressed, ordered, and compared?

When you **round**, you change the value of a number to one that is easier to work with.

Math in My World

Example 1

There are 32 people in line ahead of Cassandra to buy popcorn. About how many people is that? Round to the nearest ten.

Use a place-value chart.

1. Circle the digit to be rounded.

2. Look at the digit to its right.

hundreds	tens	ones
	③ →	2

3. If the digit is less than 5, do not change the circled digit.
   2 < 5

hundreds	tens	ones

4. Replace all the digits after the circled digit with zeros.

Use a number line.

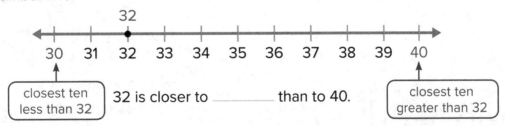

32 is closer to _____ than to 40.

Either method you use, the answer is the same. Cassandra has about _____ people in front of her.

Online Content at connectED.mcgraw-hill.com   Lesson 4 Round to the Nearest Ten 29

Example 2

Sonja sent 165 text messages on her family's cell phone. About how many messages did Sonja send?

Use a place-value chart to round 165 to the nearest ten.

1 Circle the digit to be rounded.

hundreds	tens	ones
1	⑥ →	5

2 Look at the digit to its right.

3 If the digit is 5 or greater, add 1 to the circled digit.

5 = 5

hundreds	tens	ones
1		

4 Replace all the digits after the circled digit with zeros.

So, Sonja sent about 170 text messages.

Guided Practice

Round to the nearest ten.

1.

hundreds	tens	ones
	5	8

2.

hundreds	tens	ones
	8	5

3.

hundreds	tens	ones
	7	2

Talk MATH

What should you do to round a number that ends in 5, which is exactly halfway between two numbers?

Chapter 1 Place Value

Name _____

Independent Practice

Round to the nearest ten.

4. 77 _____   **5.** 67 _____   **6.** 13 _____   **7.** 21 _____

8. 285 _____   **9.** 195 _____   **10.** 157 _____   **11.** 679 _____

12. 123 _____   **13.** 244 _____   **14.** 749 _____   **15.** 603 _____

16. 353 _____   **17.** 894 _____   **18.** 568 _____   **19.** 829 _____

Round each number to the nearest ten. Circle the row or column in which three rounded numbers are the same.

20.

37	317	35
513	766	91
251	249	245

21.

19	989	486
515	519	492
536	12	493

Round to the nearest ten. Draw a line to match each number to its rounded number.

22. 345   • 290

23. 317   • 350

24. 295   • 310

25. 291   • 320

26. 305   • 300

Lesson 4 Round to the Nearest Ten 31

Problem Solving

Processes & Practices 4 Model Math The table shows Dan's bowling scores for one week.

27. To the nearest ten, on which day was the score about 260?

28. To the nearest ten, what was Tuesday's score?

29. To the nearest ten, which day's score rounds to 250?

Monday 252
Tuesday 83
Wednesday 164
Thursday 256
Friday 290
Saturday 283
Sunday 173

Brain Builders

30. **Processes & Practices 8 Look for a Pattern** Josie is thinking of a number that when rounded to the nearest ten is 100. What could the number be? Explain.

31. **Processes & Practices 3 Find the Error** Thomas rounded the numbers below to the nearest ten. Circle his error. Explain.

184 → 180   55 → 50   295 → 300

32. **Building on the Essential Question** Why are rounded numbers easier to work with? Explain your reasoning to a friend.

32 Chapter 1 Place Value

Name _____

MY Homework

Lesson 4

Round to the Nearest Ten

Homework Helper

Need help? connectED.mcgraw-hill.com

Chase read an article about a football player who carried the ball for 437 yards. Round this number to the nearest ten.

Use a place-value chart.

1. Circle the digit to be rounded.

hundreds	tens	ones
4	③ →	7

2. Look at the digit to its right.

3. If the digit is 5 or greater, add 1 to the circled digit. If it is 4 or less, do not change it.

hundreds	tens	ones
4	4	0

4. Replace all the digits after the circled digit with zeros.

So, the football player carried the ball for about 440 yards.

Practice

Round to the nearest ten.

1. 392 _____

2. 126 _____

Use the number line to round to the nearest ten.

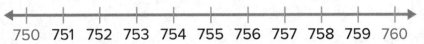

3. 753

4. 758

5. 756

Lesson 4 My Homework   33

Use the number line to round to the nearest ten.

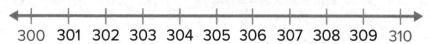

6. 302 _____   7. 304 _____   8. 305 _____

Round to the nearest ten.

9. 429 _____   10. 191 _____   11. 198 _____

Brain Builders

12. **Processes &Practices** **Use Number Sense** Kayla's dog weighs 112 pounds. Garret's dog weighs 15 pounds more than Kayla's dog. About how much does Garret's dog weigh to the nearest ten pounds?

Vocabulary Check

For each sentence, write one of the following on the line.

round   nearest ten

13. To _____ a number, you change its value to a number that is easier to work with.

14. Rounded to the _____, 167 becomes 170.

15. **Test Practice** Faith has 132 songs in her music library. She buys 11 more songs. About how many songs is that? Round to the nearest ten.

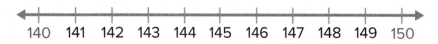

Ⓐ 130 songs   Ⓒ 145 songs

Ⓑ 140 songs   Ⓓ 150 songs

Name ....

Lesson 5
Round to the Nearest Hundred

ESSENTIAL QUESTION
How can numbers be expressed, ordered, and compared?

There can be more than one reasonable rounded number.

Math in My World

Example 1

A hydroplane is an extremely fast motor boat. In California, a record speed was set of 213 miles per hour. What is the record speed rounded to the nearest ten and rounded to the nearest hundred?

Round to the nearest ten.

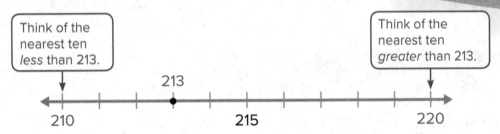

Think of the nearest ten *less* than 213.

Think of the nearest ten *greater* than 213.

Rounded to the nearest ten, the hydroplane's speed was _____ miles per hour.

Round to the nearest hundred.

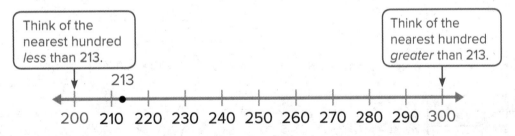

Think of the nearest hundred *less* than 213.

Think of the nearest hundred *greater* than 213.

Rounded to the nearest hundred, the hydroplane's speed was _____ miles per hour.

Example 2

Olivia placed some jellybeans in a jar. She rounded that number to the nearest hundred. To win the jar of jelly beans, the estimate needed to be guessed. What was the winning number?

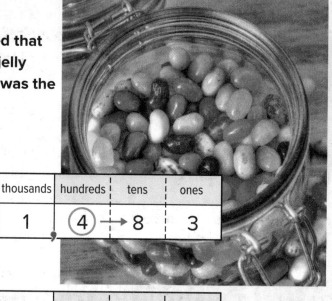

Use a place-value chart to round.

1 Circle the digit to be rounded.

2 Look at the digit to the right of the place being rounded.

3 If the digit is less than 5, do not change the circled digit. If the digit is 5 or greater, add 1 to the circled digit.

4 Replace all digits after the circled digit with zeros.

So, 1,483 rounded to the nearest hundred is 1,500.

The winning number was _____.

Helpful Hint
"Rounded" numbers have 1 or more "round" 0s at the end.

Guided Practice

Round to the nearest hundred.

1.
hundreds	tens	ones
6	2	2

_____

Round to the nearest ten and nearest hundred.

2.
hundreds	tens	ones
2	6	5

ten _____ hundred _____

Talk MATH
Is it possible for a number to be rounded to the nearest ten and to the nearest hundred and result in the same rounded number?

36 Chapter 1 Place Value

Name _____

Independent Practice

Round to the nearest hundred.

3. 750 _____   4. 1,368 _____   5. 618 _____

6. 372 _____   7. 509 _____   8. 1,216 _____

Round to the nearest ten and nearest hundred.
Cross out the rounded number that does not belong.

9. 453

   450   460   500

10. 6,333

    7,000   6,330   6,300

11. 5,037

    5,000   5,040   5,100

12. 4,776

    4,700   4,800   4,780

Round each number to the nearest hundred. Circle the row or column in which three rounded numbers are the same.

13.

113	279	367
404	321	223
189	291	363

14.

1,925	4,782	2,295
850	3,815	3,795
4,723	4,689	4,717

Circle whether each number is rounded to the nearest ten or hundred.

15. 557 rounds to 560   ten   hundred

16. 415 rounds to 400   ten   hundred

17. 89 rounds to 100   ten   hundred

18. 75 rounds to 80   ten   hundred

Lesson 5 Round to the Nearest Hundred   37

Problem Solving

19. A passenger train traveled 687 miles. To the nearest hundred, how many miles did the train travel?

_____ miles

20. Processes &Practices 2 Reason Myron has 179 postcards. He says he has about 200 cards. Did he round the number of cards to the nearest ten or nearest hundred? Explain.

Brain Builders

21. The cost for the fourth grade to take a trip to the zoo is $1,635. Each grade rounds their cost to the nearest hundred. Based on the estimates, how much more does the third grade pay? Explain your answer.

Third Grade Trip to the Zoo
$1,855

22. Processes &Practices 4 Model Math Mrs. Jones is thinking of a number that when rounded to the nearest hundred is 400. What is the number? Explain.

23. Building on the Essential Question Why might you want to round to the nearest hundred rather than the nearest ten?

38 Chapter 1 Place Value

Name _____

MY Homework

Lesson 5

Round to the Nearest Hundred

Homework Helper

Need help? connectED.mcgraw-hill.com

A roller coaster in Ohio travels a track 5,427 feet long. About how many feet is that? Round to the nearest hundred.

You can use a place-value chart.

1. Circle the digit to be rounded.

2. Look at the digit to its right.

thousands	hundreds	tens	ones
5 ,	④ →	2	7

3. If the digit is 5 or greater, add 1 to the circled digit. If it is less than 5, do not change it.

thousands	hundreds	tens	ones
5 ,	4	0	0

4. Replace all the digits after the circled digit with zeros.

So, the roller coaster track is about 5,400 feet long.

Practice

Round to the nearest hundred.

1. 688 _____

2. 4,248 _____

3. 316 _____

4. 2,781 _____

Use the number line to round to the nearest hundred.

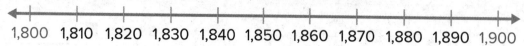

5. 1,877 _____   **6.** 1,849 _____   **7.** 1,829 _____

Round to the nearest ten and nearest hundred.

8.

hundreds	tens	ones
7	0	9

ten _____

hundred _____

9.

hundreds	tens	ones
1	8	5

ten _____

hundred _____

Problem Solving

10. Processes &Practices 2 **Reason** The table shows how many patients three doctors saw last year. To the nearest hundred, which doctor saw about 2,400 patients? Explain.

Doctor	Patients
Dr. Walters	2,493
Dr. Santos	3,205
Dr. Cooper	2,353

_____

_____

Brain Builders

11. A group of hikers plans to climb Mt. Carmy. They start at 1,270 feet. To the nearest hundred, about how many feet will they climb to the peak?

_____

12. Test Practice Mr. Miles bought a telescope that cost $3,102 and a pair of binoculars that cost $454. To the nearest hundred, how much did they cost together?

Ⓐ $3,500

Ⓑ $3,560

Ⓒ $3,600

Ⓓ $4,000

Name _____

Lesson 6
Problem-Solving Investigation
STRATEGY: Use the Four-Step Plan

ESSENTIAL QUESTION
How can numbers be expressed, ordered, and compared?

Learn the Strategy

Dina's family went to a zoo on their vacation. They learned that a roadrunner is 1 foot tall. An African elephant is 12 feet tall. How much taller is an African elephant than a roadrunner?

1 Understand
What facts do you know?

The roadrunner is _____ foot tall.

The African elephant is _____ feet tall.

What do you need to find?

how much taller an African elephant is than a _____

2 Plan

_____ the roadrunner's height from the African elephant's height.

3 Solve

```
   12    ← height of elephant
 −  1    ← height of roadrunner
 ────
```

So, the elephant is _____ feet taller than the roadrunner.

4 Check
Does your answer make sense? Explain.

_____

Online Content at connectED.mcgraw-hill.com

Lesson 6 Problem-Solving Investigation 41

Practice the Strategy

Alex wants to buy a train ticket to his grandfather's house. A one-way ticket costs $43. A round-trip ticket costs $79. Is it less expensive to buy 2 one-way tickets or one round-trip ticket?

 Understand

What facts do you know?

_____

What do you need to find?

_____

 Plan

_____

_____

 Solve

4 Check

Does your answer make sense? Explain.

_____

42 **Chapter 1** Place Value

Name

Apply the Strategy

Solve each problem by using the four-step plan.

1. The table shows the number of play tickets four friends sold on Saturday.

Louise	Malcolm
10 + 7	fourteen
Bobby	**Shelly**
20 − 5	19 + 2

   Write the number of tickets sold by each friend in standard form. Then order the numbers from *greatest* to *least*.

2. Cameron and Mara walked 2 blocks. Then they turned a corner and walked 4 blocks. If they turned around now and returned home the way they came, how many blocks will they have walked all together

Brain Builders

3. **Processes & Practices 5** **Use Math Tools** Follow the directions to find the correct height of the CN Tower in Toronto, Canada. Start with 981 feet. Subtract 200. Add 1,000. Add 4 tens, and subtract 6 ones.

4. In 1,000 years from now, what year will it be? What year will it be 100 years from now? 10 years from now? Explain your reasoning.

Lesson 6 Problem-Solving Investigation   **43**

Review the Strategies

Use any strategy to solve each problem.
- Use the four-step plan.
- Guess, check, and revise.
- Act it out.

5. The blue number cubes represent hundreds. The red number cubes represent tens. Write the standard form for each person's set of number cubes.

Robin's number cubes

Percy's number cubes

Byron's number cubes

Order these numbers from *greatest to least*.

_____

6. **Processes &Practices 1** **Plan Your Solution** Two students traveled during their summer vacation. Tina traveled 395 miles. Carl traveled 29 more miles than Tina. How many miles did Carl travel?

_____

7. **Processes &Practices 6** **Be Precise** A group of hikers traveled to the Adirondack Mountains. To the nearest ten, about how many feet did they hike if they climbed Big Slide Mountain and Algonquin Peak?

Adirondack Mountains	
Mt. Marcy	5,344 ft
Algonquin Peak	5,114 ft
Whiteface Mt.	4,867 ft
Big Slide Mt.	4,240 ft

Is there any difference in the distance if you round to the nearest hundred instead? Use <, >, or = to explain.

_____

44 Chapter 1 Place Value

Name _____

MY Homework

Lesson 6

Problem Solving: Use the Four-Step Plan

Homework Helper

Need help? connectED.mcgraw-hill.com

There were 418 tickets sold for Conrad's piano recital. About how many tickets were sold? Round to the nearest hundred.

1 Understand What facts do you know?

There were 418 tickets sold.

What do you need to find?

the number of tickets sold to the nearest hundred

2 Plan Use a number line.

3 Solve

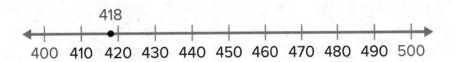

418 is closer to 400 than to 500. About 400 tickets were sold.

4 Check Check using a place-value chart. The answer is correct.

hundreds	tens	ones
④ →	1	8
4	0	0

Problem Solving

1. Victoria buys two pairs of sunglasses for $8 each. About how much did she pay for both pairs of glasses? Round to the nearest ten. Solve the problem using the four-step plan.

_____

Lesson 6 My Homework 45

Solve each problem using the four-step plan.

2. Ridgeway Elementary School collected *eight thousand, five hundred thirty-one* cans of food. Park Elementary School collected *eight thousand, six hundred forty-two* cans. Which school collected more cans of food? Write the greater number of cans in expanded form.

_____

_____

3. **Processes &Practices** **Make a Plan** There are 398 students at Rebecca's school and 462 students at Haley's school. There are 10 more students at Ronnie's school than there are at Rebecca's school. Order the number of students at each school from least to greatest.

_____

Brain Builders

4. **Processes &Practices** **Justify Conclusions** The table shows the points each player scored in a video game. Jamal's score is between the scores of Josh and Jake. Give one possible score for Jamal.

Player	Points
Josh	2,365
Jake	2,475

How did you decide on Jamal's score?

_____

_____

5. Half the distance from Los Angeles, California to New York, New York is 1,391 miles. What is the distance between the two cities rounded to the nearest hundred miles?

_____

Review

Chapter 1
Place Value

Vocabulary Check

Read each clue. Fill in the matching section of the crossword puzzle to answer each clue. Use the words in the word bank.

digit   expanded form   place value
round   standard form   word form

Across

1. A form which shows the sum of the value of the digits.

2. The usual way of writing numbers that shows only its digits.

Down

3. The value given to a digit by its place in a number.

4. The form of a number that uses written words.

5. Any symbol used to write whole numbers.

6. To find the nearest value of a number based on a given place value.

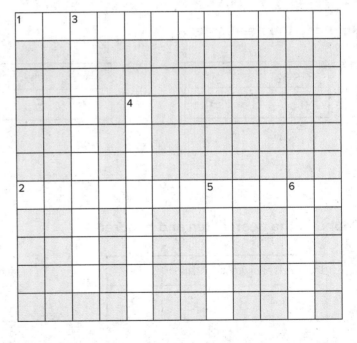

My Chapter Review 47

Concept Check

Write the highlighted digit's place and value.

7. 9**4**5 _____

8. 4,7**3**1 _____

9. **5**,409 _____

Write each number in standard form.

10. 300 + 40 + 7 _____

11. two thousand, six hundred twenty-two _____

Write 3,651 in expanded form and word form.

12. Expanded Form: _____ + _____ + _____ + _____

13. Word Form: _____

Compare. Use >, <, or =.

14. 268 ◯ 298

15. 3,499 ◯ 3,499

16. 2,675 ◯ 2,567

Round to the nearest ten.

17.
hundreds	tens	ones
4	8	4

18.
hundreds	tens	ones
2	5	9

19.
hundreds	tens	ones
7	1	2

Round to the nearest ten and hundred.

20.
thousands	hundreds	tens	ones
5	3	3	3

ten _____

hundred _____

21.
thousands	hundreds	tens	ones
2	7	8	7

ten _____

hundred _____

48 **Chapter 1** Place Value

Name _____

Problem Solving

22. Lindsey uses the digits 3, 8, 0, and 1. She uses each digit only once. Find the greatest whole number she can make.

_____

23. Alicia wrote 5,004 in word form. Find and correct her mistake.

Five hundred four

_____

_____

_____

Brain Builders

24. Keith's family bought a computer for $1,005 and a printer for $150. Margareta's family bought a computer for $1,050. Which family spent less? Explain.

_____

_____

25. Test Practice The Knights track team scored 112 points at last week's meet. This week, they scored 15 more points than last week. About how many points did The Knights team score this week? Round to the nearest ten.

Ⓐ 120 points Ⓒ 130 points

Ⓑ 127 points Ⓓ 137 points

Reflect

Chapter 1
Answering the
ESSENTIAL QUESTION

Use what you learned about place value to complete the graphic organizer.

Draw a Picture | Real-World Problem

ESSENTIAL QUESTION
How can numbers be expressed, ordered, and compared?

Vocabulary | Forms of Numbers

Now reflect on the ESSENTIAL QUESTION Write your answer below.

___

50   **Chapter 1** Place Value

Performance Task

Putting a Spin on It

Sara and De'Quan are playing a game where a spinner is spun for each player's turn. Each player receives the number of points shown on the spinner for each spin.

Show all your work to receive full credit.

Part A

To play the game, each player spins the spinner. The players take turns spinning and each player has four turns. To keep track of the scores of each player, the score at the end of each spin is rounded to the nearest ten. Complete the table by rounding the scores at the end of each player's turn to the nearest ten.

	Sara's Score	Sara's Rounded Score	De'Quan's Score	De'Quan's Rounded Score
Turn 1	78		81	
Turn 2	67		77	
Turn 3	86		72	
Turn 4	73		63	

Part B

Think about the difficulty of adding the scores that are not rounded and adding the scores that are rounded. Give a reason for why rounding the scores to the nearest ten after each turn is helpful.

Part C

Find each player's total score when they do not round after each turn and when they do round after each turn. Does rounding the scores to the nearest ten after each turn change the outcome of the game? Explain.

Part D

Sara and De'Quan decide to play the game again. The results are shown in this table. This time they decide to round their scores to the nearest ten at the very end of the game. Find each player's total score, and then round the scores to the nearest ten. Who wins the game?

	Sara	De'Quan
Turn 1	74	81
Turn 2	67	86
Turn 3	88	69
Turn 4	83	78

Part E

For the second game, would the results be different if they rounded their scores after each round instead of at the very end of the game? Explain.

Chapter 2 Addition

My Transportation

ESSENTIAL QUESTION

How can place value help me add larger numbers?

Watch a video!

Name
....

MY Chapter Project

Bake Sale

Day 1

1. Work together as a group to decide what baked item to make and sell at a bake sale.

Baked item to sell: _____

2. As a group, decide how many units of the baked item you would like to sell and the price of each unit.

Number of units to sell: _____

Unit price: _____

3. Determine the amount of money your group will make if you sell all the items at the unit price.

Total amount: _____

4. In the space below, sketch a design for your item that could be used on a sign at the bake sale. Be sure to include the name of the baked item, a picture, and the unit price.

52 Chapter 2 Addition

Name _____

Am I Ready?

Add.

1. 5
   +4

2. 6
   +7

3. 9
   +6

4. 4
   +8

5. 9 + 2 = _____

6. 4 + 6 = _____

7. 9 + 8 = _____

8. 7 + 7 = _____

9.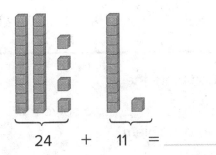

24 + 11 = _____

10.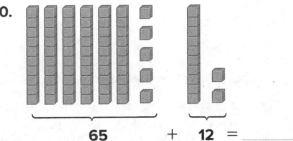

65 + 12 = _____

11. What number is 10 more than 66?

   _____

12. What number is 100 more than 800?

   _____

Round to the nearest ten.

13. 72 _____

14. 17 _____

15. 63 _____

16. 88 _____

Round to the nearest hundred.

17. 470 _____

18. 771 _____

19. 301 _____

20. 149 _____

Shade the boxes to show the problems you answered correctly.

How Did I Do? | 1 | 2 | 3 | 4 | 5 | 6 | 7 | 8 | 9 | 10 | 11 | 12 | 13 | 14 | 15 | 16 | 17 | 18 | 19 | 20 |

Online Content at connectED.mcgraw-hill.com

Name ....

MY Math Words

Review Vocabulary

addend   addition sentence   sum

Making Connections
Use the review vocabulary to complete the graphic organizer. You will use numbers in some of your answers.

Bella and Tyler looked for birds at the park. Bella saw 10 tundra swans. Tyler saw 30 cardinals. How many birds did they see all together?

Complete the _____ _____ to solve the word problem.

10 + _____ = _____

The _____ are 10 and 30.

The _____ of 10 and 30 is 40.

54 Chapter 2 Addition

MY Vocabulary Cards

Lesson 2-1

Associative Property of Addition

$(2 + 5) + 1 = 2 + (5 + 1)$

Lesson 2-8

bar diagram

|←------- ? $ spent -----------→|
| $3,295 | $3,999 |
| last year | this year |

Lesson 2-1

Commutative Property of Addition

$12 + 15 = 15 + 12$

Lesson 2-5

estimate

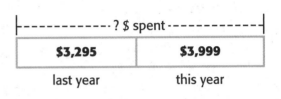

about $30

Lesson 2-1

Identity Property of Addition

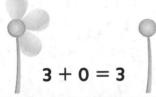

$3 + 0 = 3$

Lesson 2-1

mental math

$5 + 7 + 5 = \blacksquare$
$10 + 7 = 17$

Lesson 2-1

parentheses

$(3 + 4) + 2 = 3 + (4 + 2)$

Lesson 2-2

pattern

+	0	1	2	3
0	0	1	2	3
1	1	2	3	4
2	2	3	4	5
3	3	4	5	6

Ideas for Use

- During this school year, create a separate stack of cards for key math verbs, such as *regroup*. These verbs will help you in your problem solving.

- Use a blank card to write this chapter's essential question. Use the back of the card to write or draw examples that help you answer the question.

A bar diagram is used to illustrate number relationships.

Jimena made $1,595 last year and $1,876 this year. Draw a bar diagram to show this problem.

The property which states that the grouping of addends does not change the sum.

Write your own example to show this property.

_____

_____

A number close to an exact value.

Describe when you might need an estimate.

_____

_____

The order in which two numbers are added does not change the sum.

Complete the number sentence to show the Commutative Property of Addition.

$11 + 7 = $ _____

Ordering or grouping numbers so they are easier to add in your head.

When might it be important to use mental math at a store?

_____

_____

If you add zero to a number, the sum is the same as the given number.

Replace the suffix *-ity* in *identity* with a new suffix. Use it in a sentence.

_____

_____

A set of numbers that follows a certain order.

Write your own pattern using numbers.

_____

Symbols that are used to group numbers. They show how to group operations in a number sentence.

Use a dictionary to find the singular form of *parentheses*.

_____

MY Vocabulary Cards

Processes & Practices

Lesson 2-6

reasonable

$$682 \rightarrow 680$$
$$\underline{+\ 17} \rightarrow \underline{+\ 20}$$
$$699 \quad \boxed{700}$$

700 is a reasonable estimate.

Lesson 2-6

regroup

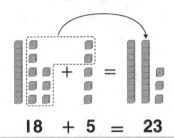

$$18 + 5 = 23$$

Lesson 2-7

unknown

$$21 + 6 = \blacksquare$$

↑ unknown

Ideas for Use

- During this school year, create a separate stack of cards for key math verbs, such as regroup. These verbs will help you in your problem solving.

- Use a blank card to write this chapter's essential question. Use the back of the card to write or draw examples that help you answer the question.

To use place value to exchange equal amounts when renaming a number.

What is the prefix in *regroup*? What does it mean?

Within the bounds of making sense.

The prefix un- can mean "not." What does the word *unreasonable* mean?

A missing number, or the number to be solved for.

Write an example of an unknown at the beginning of a number sentence.

MY Foldable

FOLDABLES Follow the steps on the back to make your Foldable.

Thousands	Hundreds	Tens	Ones
1,000	100	10	1
1,000	100	10	1
1,000	100	10	1
1,000	100	10	1
1,000	100	10	1
1,000	100	10	1
1,000	100	10	1
1,000	100	10	1
1,000	100	10	1
1,000	100	10	1

 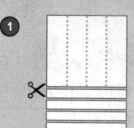 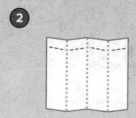 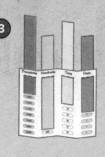

Name _____

Lesson 1
Addition Properties

ESSENTIAL QUESTION
How can place value help me add larger numbers?

In math, properties are rules you can use with numbers.

 Math in My World

Example 1

Sid has 4 blue pens and 5 red pens. Mario has 5 blue pens and 4 red pens. How many pens does each boy have?

Find 4 + 5. Then find 5 + 4.

4 + 5 = _____ .   Sid has _____ pens.

5 + 4 = _____ .   Mario has _____ pens.

This shows the **Commutative Property of Addition**. The order in which the numbers were added did not change the sum.

Example 2

Anna practiced the piano for 3 hours on Friday. She did not practice at all on Saturday. How many hours did she practice all together?

Find 3 + 0. Then find 0 + 3.

3 + 0 = _____ ← Anna practiced 3 hours all together.

0 + 3 = _____ ←

This shows the **Identity Property of Addition**. The sum of any number and zero is the number.

Online Content at connectED.mcgraw-hill.com   Lesson 1 Addition Properties **61**

Sometimes you may want to group numbers in a way that makes them easier to add in your head. This is **mental math**.

Example 3

Matthew saw 9 sailboats, 4 rowboats, and 6 canoes on the lake. How many boats did he see all together?

Find (9 + 4) + 6.

> **Parentheses** show how to group operations. These parentheses tell you to add 9 + 4 first.

Helpful Hint
9 + 4 + 6 = 19 is a number sentence because it contains numbers, operations, and an equals sign.

Since 4 + 6 = 10, finding 4 + 6 is easier than finding 9 + 4. The way in which the addends are grouped does not change the sum. This is the **Associative Property of Addition**.

(9 + 4) + 6 = 9 + (4 + 6)   Associative Property of Addition

= 9 + 10   Add 4 + 6.

= 19   Add 9 and 10.

So, (9 + 4) + 6 = _____. There are _____ boats on the lake.

Guided Practice

Find each sum. Draw a line to the correct addition property.

1. 6 + 5 = _____

   5 + 6 = _____

    • Identity

2. (5 + 7) + 3 = _____

   5 + (7 + 3) = _____

    • Associative

Talk MATH

How can you use the Associative Property to add 7, 8, and 3?

3. 0 + 12 = _____

    • Commutative

62 **Chapter 2** Addition

Name _____

Independent Practice

Find each sum. Draw a line to the correct addition property.

4. (2 + 5) + 8 = _____

   2 + (5 + 8) = _____

   • Commutative Property

5. 2 + 8 = _____

   8 + 2 = _____

6. 9 + 2 = _____

   2 + 9 = _____

   • Associative Property

7. 100 + 0 = _____

8. 4 + (6 + 3) = _____

   (4 + 6) + 3 = _____

   • Identity Property

Algebra Use an addition property to complete.

9. 6 + _____ = 6

10. (7 + 9) + _____ = (9 + 7) + 3

11. 9 + 2 = 2 + _____

12. (8 + 3) + _____ = 8 + (3 + 2)

Find each sum mentally.

13. (7 + 1) + 9 = _____

14. (7 + 5) + 5 = _____

15. Complete the number sentence for the figures below which shows the Associative Property of Addition.

(3 + 5) + _____ = 3 + (_____ + _____)

Lesson 1 Addition Properties 63

Problem Solving

For Exercises 16–17, write a number sentence and identify the property.

16. In three baseball games, the Tigers scored 7, 4, and 6 runs. How many runs did the Tigers score in all 3 games?

(_____ + 4) + 7 = _____

_____ Property of Addition

17. Processes &Practices 4 **Model Math** Kelly collected 16 vacation brochures last summer. This summer she did not collect any brochures. How many brochures did she collect all together?

Brain Builders

18. Mrs. Jackson bought 27 blue notebooks and some red notebooks. Mr. Mendez bought 19 blue notebooks and 27 red notebooks. They bought the same number of notebooks. How many red notebooks did Mrs. Jackson buy? Explain how you know.

19. Processes &Practices 2 **Reason** Can the Commutative Property be used for subtraction? Give an example to justify your answer.

20. **Building on the Essential Question** How can addition properties help me add whole numbers? Give an example.

64 Chapter 2 Addition

Name _____

MY Homework

Lesson 1
Addition Properties

Homework Helper

Need help? connectED.mcgraw-hill.com

Jaycee bought 3 yellow apples and 4 green apples.
Marshall bought 4 yellow apples and 3 green apples.
How many apples did each person buy?

$3 + 4 = 7$ $4 + 3 = 7$ ← Commutative Property of Addition

They each bought 7 apples. The order in which the addends are added does not change the sum.

You learned two other addition properties.

The Identity Property shows the sum of any number and zero is the number.

$7 + 0 = 7$

The Associative Property shows the way the addends are grouped does not change the sum.

$(7 + 4) + 3 = 7 + (4 + 3)$
$= 7 + 7$
$= 14$

Practice

Draw lines to match each property of addition with its correct example.

1. $3 + 4 = 7$ $4 + 3 = 7$ • Commutative Property of Addition

2. $7 + 0 = 7$ • Associative Property of Addition

3. $7 + (3 + 4) = (7 + 3) + 4$ • Identity Property of Addition

Lesson 1 My Homework 65

Find each sum. Identify the addition property.

4. 46 + 0 = _____

5. (7 + 9) + 3 = _____   7 + (9 + 3) = _____

_____ Property

_____ Property

Brain Builders

Find each sum mentally. Tell someone how you used a property.

6. (6 + 8) + 2 = _____

7. 3 + (2 + 4) = _____

8. **Processes &Practices** 1 **Make Sense of Problems** Pedro and Jamal collected yellow and red leaves. They each collected the same total number of leaves. How many red leaves did Jamal collect? Tell someone how you used a property. Write an equation to support your answer.

_____

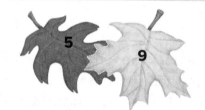

Pedro's leaves

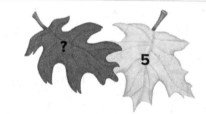

Jamal's leaves

Vocabulary Check

Draw a line to match the vocabulary word(s) to its example.

9. Commutative Property   • symbols which show grouping

10. parentheses   • (3 + 1) + 4 = 8   3 + (1 + 4) = 8

11. Identity Property   • 5 + 6 = 11   6 + 5 = 11

12. Associative Property   • 2 + 0 = 2

13. **Test Practice** Which number sentence is an example of the Associative Property?

Ⓐ 5 + 1 = 3 + 3

Ⓒ (8 + 2) + 5 = 8 + (2 + 5)

Ⓑ 583 + 0 = 583

Ⓓ 3 + 5 = 5 + 3

Name ....

Lesson 2
Patterns in the Addition Table

ESSENTIAL QUESTION
How can place value help me add larger numbers?

Study the addition table for number **patterns**. Look for sets of numbers that follow a certain order.

Make a pattern!

Math in My World

Example 1

Danny colored a pattern of squares from left to right on the downward diagonal in yellow. Describe the pattern.

Even Numbers

Finish Danny's pattern of even numbers. Color the squares.

0, 2, 4, 6, 8, 10, 12, 14, 16, 18, 20

There is a pattern of add _____ .

When you add _____ to an even number, the sum is an _____ number.

Odd Numbers

Start with the green square. Color the pattern of odd numbers on the downward diagonal in green. Write the numbers.

columns of addends

+	0	1	2	3	4	5	6	7	8	9	10
0	0	1	2	3	4	5	6	7	8	9	10
1	1	2	3	4	5	6	7	8	9	10	11
2	2	3	4	5	6	7	8	9	10	11	12
3	3	4	5	6	7	8	9	10	11	12	13
4	4	5	6	7	8	9	10	11	12	13	14
5	5	6	7	8	9	10	11	12	13	14	15
6	6	7	8	9	10	11	12	13	14	15	16
7	7	8	9	10	11	12	13	14	15	16	17
8	8	9	10	11	12	13	14	15	16	17	18
9	9	10	11	12	13	14	15	16	17	18	19
10	10	11	12	13	14	15	16	17	18	19	20

rows of addends

1, _____, _____, _____, _____, _____, _____, _____, _____,

There is a pattern of add _____ . When you add _____ to an odd number, the sum is an _____ number.

Online Content at connectED.mcgraw-hill.com

Lesson 2 Patterns in the Addition Table **67**

Example 2

1. What pattern of numbers do you see in the diagonal of yellow boxes?

_____

2. Look at the circled sum. Follow left and above to the circled addends.

3. Draw a triangle around the sum in the addition table that has the same addends. Follow left and above to its addends. Complete the number sentences.

+	0	1	2	3	4	5	6	7	8	9	10
0	0	1	2	3	4	5	6	7	8	9	10
1	1	2	3	4	5	6	7	8	9	10	11
2	2	3	4	5	6	7	8	9	10	11	12
3	3	4	5	6	7	8	9	10	11	12	13
4	4	5	6	7	8	9	10	11	12	13	14
5	5	6	7	8	9	10	11	12	13	14	15
6	6	7	8	9	10	11	12	13	14	15	16
7	7	8	9	10	11	12	13	14	15	16	17
8	8	9	10	11	12	13	14	15	16	17	18
9	9	10	11	12	13	14	15	16	17	18	19
10	10	11	12	13	14	15	16	17	18	19	20

addends   sum

____ + ____ = 7

____ + ____ = 7

The two number sentences are an example of the _____ Property.

Guided Practice

Describe Danny's new pattern in the addition table below.

1. When _____ is added to a number, the sum is that number.

2. This is an example of the _____ Property of Addition.

+	0	1	2	3	4	5	6
0	0	1	2	3	4	5	6
1	1	2	3	4	5	6	7
2	2	3	4	5	6	7	8
3	3	4	5	6	7	8	9
4	4	5	6	7	8	9	10
5	5	6	7	8	9	10	11
6	6	7	8	9	10	11	12

Talk MATH

How do you find patterns in numbers?

Name _____

Independent Practice

Use the addition table.

+	0	1	2	3	4	5	6	7	8	9	10
0	0	1	2	3	4	5	6	7	8	9	10
1	1	2	3	4	5	6	7	8	9	10	11
2	2	3	4	5	6	7	8	9	10	11	12
3	3	4	5	6	7	8	9	10	11	12	13
4	4	5	6	7	8	9	10	11	12	13	14
5	5	6	7	8	9	10	11	12	13	14	15
6	6	7	8	9	10	11	12	13	14	15	16
7	7	8	9	10	11	12	13	14	15	16	17
8	8	9	10	11	12	13	14	15	16	17	18
9	9	10	11	12	13	14	15	16	17	18	19
10	10	11	12	13	14	15	16	17	18	19	20

3. Shade a diagonal of numbers **blue** that show the sums equal to 8.

4. Shade a diagonal of numbers **green** that show the sums equal to 5.

5. Shade a row of numbers **yellow** that represent sums with one addend of 4.

6. Shade a column of numbers **pink** that represent sums with one addend of 6.

7. Shade two squares **purple** that each represent the sum of 3 and 9. What property does this show?

   _____

8. Circle two squares that each represent a sum of 0 and 10. What two properties does this show?

   _____
   _____

9. Shade two addends **red** that have a sum of 8. Complete the number sentence. Write the greater addend first.

   _____ + _____ = 8

   Use the Commutative Property of Addition and shade the other two addends **red**. Complete the other addition sentence.

   _____ + _____ = 8

Lesson 2 Patterns in the Addition Table   69

Problem Solving

Use the addition table.

10. **Processes &Practices** 5 **Use Math Tools** Marlo stacked 8 boxes. She had no more boxes to stack. Find the total number of boxes stacked. Shade two squares that each represent the sum. Write two number sentences.

+	0	1	2	3	4	5	6	7	8	9	10
0	0	1	2	3	4	5	6	7	8	9	10
1	1	2	3	4	5	6	7	8	9	10	11
2	2	3	4	5	6	7	8	9	10	11	12
3	3	4	5	6	7	8	9	10	11	12	13
4	4	5	6	7	8	9	10	11	12	13	14
5	5	6	7	8	9	10	11	12	13	14	15
6	6	7	8	9	10	11	12	13	14	15	16
7	7	8	9	10	11	12	13	14	15	16	17
8	8	9	10	11	12	13	14	15	16	17	18
9	9	10	11	12	13	14	15	16	17	18	19
10	10	11	12	13	14	15	16	17	18	19	20

What two properties does this show?

_____

11. Pedro ran 3 miles on Sunday and 2 miles on Monday. Find the total number of miles he ran. Shade two squares that each represent the sum. Write two number sentences.

_____

What property does this show?

Brain Builders

12. **Processes &Practices** 4 **Model Math** Write a real-world problem for which you can use the addition table and the Commutative Property of Addition to solve. Then solve.

_____

_____

13. **Building on the Essential Question** How can addition patterns help me add mentally? How can they help me check my work?

_____

_____

Chapter 2 Addition

Name ....

MY Homework

Lesson 2

Patterns in the Addition Table

Homework Helper

Need help? connectED.mcgraw-hill.com

Eddie colored the top row of an addition table blue. What is the pattern?

The sum of any number and zero is the number. This shows the Identity Property of Addition.

Using green , Eddie started at 2 and colored a downward diagonal pattern. What is the pattern?

Adding 2 to an even number shows a pattern of even numbers.

Using purple , Eddie started at 5 and colored a downward diagonal. What is the pattern?

Adding 2 to an odd number shows a pattern of odd numbers.

+	0	1	2	3	4	5	6	7	8	9	10
0	0	1	2	3	4	5	6	7	8	9	10
1	1	2	3	4	5	6	7	8	9	10	11
2	2	3	4	5	6	7	8	9	10	11	12
3	3	4	5	6	7	8	9	10	11	12	13
4	4	5	6	7	8	9	10	11	12	13	14
5	5	6	7	8	9	10	11	12	13	14	15
6	6	7	8	9	10	11	12	13	14	15	16
7	7	8	9	10	11	12	13	14	15	16	17
8	8	9	10	11	12	13	14	15	16	17	18
9	9	10	11	12	13	14	15	16	17	18	19
10	10	11	12	13	14	15	16	17	18	19	20

Practice

Use the addition table above.

1. Shade a diagonal of odd numbers red .

2. Shade a diagonal of even numbers yellow .

Lesson 2 Patterns in the Addition Table 71

Use the addition table.

3. Circle two squares that each represent a sum of 3 and 4. This shows the Commutative Property of Addition.

   3 + 4 = _____ and 4 + 3 = _____

4. Circle the two addends which make a sum of the shaded 12. Write the number sentence.

   _____

5. Shade a diagonal of numbers **green** that show the sums equal to 9.

+	0	1	2	3	4	5	6	7	8	9	10
0	0	1	2	3	4	5	6	7	8	9	10
1	1	2	3	4	5	6	7	8	9	10	11
2	2	3	4	5	6	7	8	9	10	11	12
3	3	4	5	6	7	8	9	10	11	12	13
4	4	5	6	7	8	9	10	11	12	13	14
5	5	6	7	8	9	10	11	12	13	14	15
6	6	7	8	9	10	11	12	13	14	15	16
7	7	8	9	10	11	12	13	14	15	16	17
8	8	9	10	11	12	13	14	15	16	17	18
9	9	10	11	12	13	14	15	16	17	18	19
10	10	11	12	13	14	15	16	17	18	19	20

Brain Builders

6. Shade a row of numbers **yellow** that represents sums with one addend of 10. Describe the pattern that you see.

   _____

7. **Processes &Practices** ➌ **Justify Conclusions** Jasmine had 11 friends over at her house. Every time the doorbell rang, 2 more friends arrived. The doorbell rang 3 times. How many friends did Jasmine have over all together? Is your answer even or odd? Explain.

   _____

   _____

8. Steve colors the following sums on an addition table. Describe the pattern. If he continues the pattern, will the numbers continue to be even? Explain.

   12, 14, 16, 18

   _____

   _____

9. **Test Practice** Danielle is saving for a bicycle. Her bank deposits for the last 4 weeks are shown in the table. If the pattern continues, how much will her bank deposit be in 2 more weeks?

   Ⓐ $2   Ⓑ $17   Ⓒ $19   Ⓓ $21

Bank Deposits
$9
$11
$13
$15
?
?

72 Need more practice? Download Extra Practice at connectED.mcgraw-hill.com

Name _____

Lesson 3
Addition Patterns

ESSENTIAL QUESTION
How can place value help me add larger numbers?

Math in My World

Example 1

The bank book shows how much money was added to Bart's account each time he visited the bank. How much money did Bart have after each trip to the bank? Complete Bart's bank book.

	Bart's Bank Book			
	$		5	7	5
$1 more	$		5	7	
$100 more	$			7	6
$1,000 more	$		, 6	7	6

← 1st trip →
← 2nd trip →
← 3rd trip →

thousands	hundreds	tens	ones
	5	7	5
	5	7	ⓖ
	⑥	7	6
①,	6	7	6

The circled digits show which place's value changed each time.

So, Bart had $ _____ after the first trip, $ _____ after the second trip, and $ _____ after the third trip.

Bart added some money to the $1,676 he already had in his account. He now has $1,686. Complete the number sentence to show how much he added.

$1,676 + _____ = $1,686

Online Content at connectED.mcgraw-hill.com

Lesson 3 Addition Patterns 73

Example 2

Patrick kept track of the miles his family traveled on a trip. Each time they stopped, he wrote down the miles from the odometer. Patrick noticed a pattern in the numbers he wrote. Describe the pattern.

+ 100

Each time they stopped, the numbers increased by _____ miles.

Write the next number in the pattern above.

So, Patrick's family stopped every _____ miles.

Guided Practice

Write the number in the place-value chart.

1. 100 more than 3,728

thousands	hundreds	tens	ones
3	7	2	8

2. 1 more than 281

hundreds	tens	ones
2	8	1

3. Complete the number sentence.

thousands	hundreds	tens	ones
6	3	2	5
6	4	2	5

6,325 + _____ = 6,425

Talk MATH

Tell what happens to the digits in the number 1,057 if 100 is added to that number.

74 Chapter 2 Addition

Name _____

Independent Practice

Write the number.

4. 1 more than 972

5. 1,000 more than 374

6. 10 more than 310

7. 1,000 more than 8,993

8. 10 more than 1,437

9. 100 more than 2,819

10. 100 more than 173

11. 10 more than 6,910

Complete the number sentence.

12. 974 + _____ = 975

13. 1,234 + _____ = 2,234

14. 8,264 + _____ = 9,264

15. 1,038 + _____ = 1,138

16. 6,123 + _____ = 6,223

17. 8,877 + _____ = 8,887

Identify and complete the number pattern.

18. 6,282; 7,282; _____ ; 9,282

    The number pattern is _____ .

19. 9,379; _____ ; 9,381; 9,382

    The number pattern is _____ .

20. 7,874; 7,884; _____ ; 7,904; _____ ; _____

    The number pattern is _____ .

21. Add to move up the stairs.

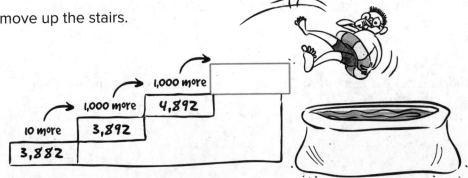

Lesson 3 Addition Patterns 75

Problem Solving

22. **Processes &Practices 8** **Look for a Pattern** A factory packages one bag of balloons each second. Each balloon below represents the total number of balloons packaged after 1 more second. How many balloons are in each bag?

Complete the pattern.

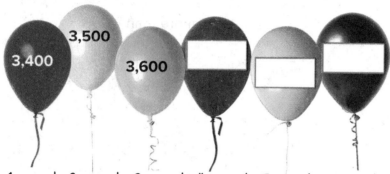

1 second 2 seconds 3 seconds 4 seconds 5 seconds 6 seconds

23. It takes one second to fill a carton with bags of balloons. Each number below represents the total number of bags of balloons packaged into cartons after 1 more second. Complete the pattern.

    4,720; 4,730; 4,740; _____ ; _____ ; _____

Brain Builders

24. **Processes &Practices 2** **Use Number Sense** Laurel is playing a hopscotch game. Help her fill in the empty hopscotch spaces by writing the number that is 1, 10, 100, or 1,000 more according to the pattern.

Start	4,500	4,600		4,800					
					5,801	5,802		5,804	5,805
6,804	6,803			6,800					

For the last space, write a number sentence to support your answer.

_____

25. **Building on the Essential Question** How does place value help me add mentally?

_____

_____

76 Chapter 2 Addition

Name ....

MY Homework

Lesson 3
Addition Patterns

Homework Helper

Need help? connectED.mcgraw-hill.com

Over several years, Mrs. Bowers' students made a total of 2,367 paper cranes. This year, she challenged her students to make the new total 2,467 paper cranes. How many paper cranes do they need to make this year to meet the challenge?

Use a place-value chart to find which place changed in value.

The new number is 100 more.

2,367 + 100 = 2,467

thousands	hundreds	tens	ones
2	③	6	7
2	④	6	7

So, this year's class needs to make 100 paper cranes.

Practice

Write the number in the place-value chart.

1. 10 more than 567

hundreds	tens	ones
5	6	7

2. 1 more than 358

hundreds	tens	ones
3	5	8

3. 1,000 more than 1,529

thousands	hundreds	tens	ones
1	5	2	9

4. 100 more than 5,834

thousands	hundreds	tens	ones
5	8	3	4

Lesson 3 My Homework 77

Complete the number sentence.

5.
thousands	hundreds	tens	ones
1	2	7	1
2	2	7	1

1,271 + _____ = 2,271

6.
thousands	hundreds	tens	ones
4	2	4	4
4	3	4	4

4,244 + _____ = 4,344

Write the number.

7. 10 more than 1,465

8. 100 more than 8,699

9. 1,000 more than 3,007

Brain Builders

Identify and complete the number pattern. What would the next number be?

10. 2,378; 2,478; 2,578; _____ ; _____ ; 2,878; 2,978

The number pattern is _____ .

11. 5,903; 5,913; 5,923; _____ ; _____ ; 5,953; 5,963

The number pattern is _____ .

Problem Solving

12. **Processes &Practices 4** Model Math A baby horse weighed 104 pounds at birth. It gained 100 pounds each month. How much does the horse weigh in 4 months? Write a number sentence to support your answer.

13. **Test Practice** Which pattern shows 100 more? Explain how you know.

Ⓐ 1,456; 1,556; 1,656; 1,756

Ⓑ 4,987; 4,887; 4,787; 4,687

Ⓒ 5,832; 5,833; 5,834; 5,835

Ⓓ 6,001; 7,001; 8,001; 9,001

78 Need more practice? Download Extra Practice at connectED.mcgraw-hill.com

Name _____

Lesson 4
Add Mentally

ESSENTIAL QUESTION
How can place value help me add larger numbers?

  Math in My World

Example 1

151 seats   128 seats

How many seats are in the two train cars?

Find 151 + 128.

$$151 = 100 + 50 + 1$$
$$+\ 128 = \square + \square + \square$$
$$\square + \square + \square = \square$$

↑ Add the hundreds.   ↑ Add the tens.   ↑ Add the ones.

Helpful Hint
Expanded form gives the value of each digit in a number.

So, 151 + 128 = _____. There are _____ seats in the two train cars.

Example 2

It is easy to make 100 from numbers that end in 98 or 99.

Find 134 + 99.

$$\underset{-1}{134} + \underset{+1}{99}$$

$$\square + \square \quad \text{These numbers are easier to add.}$$

$$133 + 100 = \square$$

So, 134 + 99 = 233.

Online Content at connectED.mcgraw-hill.com

Lesson 4 Add Mentally 79

Make either addend a ten such as 10, 20, 30, and so on.
These numbers are easier to add mentally.

Example 3

There are 37 students from Grade 3A and 25 students from Grade 3B on the bus. How many students are on the bus? Use mental math to find 37 + 25.

One Way Change 25 to 30.

37 − 5 ← Take or subtract 5 from the first addend. → ☐

+ 25 + 5 ← Give or add 5 to the second addend. → ☐

32 and 30 are easier numbers to add.

So, ____ + ____ = ____. There are ____ students on the bus.

Another Way Change 37 to 40.

37 + 3 ← Give or add 3 to the first addend. → ☐

+ 25 − 3 ← Take or subtract 3 from the second addend. → ☐

40 and 22 are easier numbers to add.

So, ____ + ____ = ____. There are ____ students on the bus.

Guided Practice

1. Break apart the addends to add mentally.

    79 = 70 + ☐
    + 54 = 50 + ☐
    ─────────────
    ☐ = ☐ + ☐

2. Make a ten to add mentally.

    64 + 8
    − 2   + 2
    ─────────
    ☐ + ☐ = ☐

Would you rather make a ten or a hundred when finding 156 + 262? Explain.

80 Chapter 2 Addition

Name

Independent Practice

Break apart the addends to add mentally.

3. $46 = \square + \square$
    $+53 = 50 + 3$
    ―――――――――――
    $99 = \square + 9$

4. $67 = \square + 7$
    $+12 = 10 + \square$
    ―――――――――――
    $\square = \square + \square$

5. $63 = 60 + \square$
    $+24 = \square + \square$
    ―――――――――――
    $\square = 80 + \square$

6. $325 = \square + \square + \square$
    $+625 = \square + \square + \square$
    ―――――――――――
    $\square = \square + \square + \square$

Make a ten or a hundred to add mentally.

7. $47 + 99$
    $-1 \quad +1$
    ―――――――
    $\square + \square = \square$

8. $31 + 299$
    $-1 \quad +1$
    ―――――――
    $\square + \square = \square$

9. $447 + 123$
    $+3 \quad -3$
    ―――――――
    $\square + \square = \square$

10. $539 + 356$
    $\square \quad \square$
    ―――――――
    $540 + \square = \square$

11. $127 + 145$
    $\square \quad \square$
    ―――――――
    $\square + \square = \square$

12. $799 + 134$
    $\square \quad \square$
    ―――――――
    $\square + \square = \square$

Lesson 4 Add Mentally 81

Problem Solving

13. Sylvia is finding 135 + 456. Explain how she can find the sum mentally.

_____

_____

14. Processes &Practices 6 **Explain to a Friend** There are 49 seats in the balcony of the theater. Use a mental math strategy to find the total number of seats in the balcony and on the main floor of the theater.

73 seats main floor

_____

Brain Builders

15. Use mental math to find the total amount of money Yolanda will spend when she buys the following items. Explain your strategy to a friend.

$9 $5 $11 $29

16. Processes &Practices 3 **Find the Error** Carmine mentally found the sum to 56 + 36. Find his mistake and correct it.

56 + 36
+ 4
―――――
60 + 36 = 96

_____

_____

17. **Building on the Essential Question** Why are some numbers easier to add than others? Give an example.

_____

_____

_____

Name ....

Lesson 4
Add Mentally

Homework Helper

Need help? connectED.mcgraw-hill.com

There are 58 passengers on a subway train. At the next stop, 33 more passengers board the train. How many passengers are there all together?

You can add mentally.

Make one addend a ten such as 10, 20, 30, and so on.

58 + 2 ← Give, or add, 2 to the first addend. → 60
+ 33 − 2 ← Take, or subtract, 2 from the second addend. → + 31
————————————————————————————————
                                                        91

Helpful Hint
60 and 31 are easier numbers to add.

So, 60 + 31 = 91. There are 91 passengers.

Practice

Break apart the addends to add mentally.

1. 41 = 40 + ☐
  + 26 = ☐ + 6
  ——————————————
    ☐ = 60 + 7

2. 328 = ☐ + 20 + 8
  + 254 = 200 + ☐ + ☐
  ——————————————
    ☐ = ☐ + ☐ + 12

Make a ten or a hundred to add mentally.

3. 76 + 7
    −3   +3
    ☐ + ☐ = ☐

4. 598 + 256
    +2   −2
    ☐ + ☐ = ☐

Lesson 4 My Homework 83

Make a ten or a hundred to add mentally.

5. 339 + 123
   ☐   ☐
   ☐ + ☐ = ☐

6. 399 + 428
   ☐   ☐
   ☐ + ☐ = ☐

Brain Builders

Break apart the addends to add mentally. Explain to a friend the strategy you used.

7. 767 = ☐ + ☐ + ☐
   + 29 = ☐ + ☐
   ☐ = ☐ + ☐ + ☐

8. 214 = ☐ + ☐ + ☐
   + 127 = ☐ + ☐ + ☐
   ☐ = ☐ + ☐ + ☐

9. **Processes & Practices 2** **Use Number Sense** Lambert's Livery rents canoes every weekend. How many total canoes were rented in June and July? Explain how you used mental math to find the sum.

Lambert's Livery Canoe Rentals	
Month	Rentals
June	154
July	198
August	176

10. **Test Practice** During Science Week, there were 77 visitors on Monday and 28 visitors on Tuesday. During Math Week, there were 10 more visitors than during Science Week. How many visited during Math Week?

    Ⓐ 49 visitors   Ⓒ 115 visitors

    Ⓑ 105 visitors   Ⓓ 215 visitors

Check My Progress

Vocabulary Check

Choose the correct word(s) to complete the sentence.

Associative   Commutative   Identity

mental math   parentheses   pattern

1. The _____ Property of Addition states that the order in which addends are added does not change the sum.

2. A set of numbers that follow a certain order is a _____.

3. You can use _____ to add numbers in your head.

4. The _____ Property of Addition states that the way addends are grouped does not change the sum.

5. _____ show which numbers to add first.

6. The _____ Property of Addition states that the sum of any number and zero is that number.

Concept Check

Make a ten or a hundred to add mentally.

7. 99 + 46
    ☐   ☐
    ☐ + ☐ = ☐

8. 641 + 199
    ☐   ☐
    ☐ + ☐ = ☐

Break apart the addends to add mentally.

9. $256 = 200 + 50 + \boxed{}$
    $+125 = \boxed{} + 20 + 5$

    $\boxed{} = \boxed{} + \boxed{} + \boxed{}$

Identify and complete the number pattern.

10. 573; 673; _____ ; 873

    The number pattern is _____ .

11. 2,930; _____ ; 2,950; 2,960

    The number pattern is _____ .

Brain Builders

Write the number in the place-value chart. Explain to a friend how to use place-value to find the number.

12. 1,000 more than 2,491

thousands	hundreds	tens	ones
2	4	9	1

13. 100 more than 8,762

thousands	hundreds	tens	ones
8	7	6	2

14. At a book sale, Emily buys 4 books. Paul buys 8 books. Genie buys 6 books. Show an addition property to find how many books Emily, Paul, and Genie buy all together. Write the total number of books and the addition property you used.

_____

15. **Test Practice** On Monday, Lisa does 9 push ups. On Tuesday, she does 0 push ups. Which addition property can be used to find the number of push ups Lisa does all together?

    Ⓐ Identity Property   Ⓒ Associative Property

    Ⓑ Commutative Property   Ⓓ Pattern Property

Name _____

Lesson 5
Estimate Sums

ESSENTIAL QUESTION
How can place value help me add larger numbers?

When you do not need an exact number, a word such as *about*, is sometimes used. You can find an estimate instead. An **estimate** is a number close to the exact number.

 Math in My World

Example 1

The Board Shop sold 342 snowboards and 637 pairs of boots in the last year. About how many snowboards and pairs of boots did they sell all together?

Estimate 342 + 637.
Round, and then add.
When estimating, you can round to the nearest ten or hundred.

One Way Hundreds Place

hundreds	tens	ones
3	4	2
+ 6	3	7

→ ☐
+ ☐
———
☐

To the nearest hundred, the Board Shop sold about _____ snowboards and pairs of boots.

Another Way Tens Place

hundreds	tens	ones
3	4	2
+ 6	3	7

→ ☐
+ ☐
———
☐

To the nearest ten, the Board Shop sold about _____ snowboards and pairs of boots.

You can have more than one good estimate.

Example 2

A board game has $4,140 in play money. About how much money would there be if two games were put together?

Estimate $4,140 + $4,140.
Round to the hundreds place.

$4,140 ⟶ $4,100
+ $4,140 ⟶ + ☐
―――――   ―――――
              ☐

So, the two games have about _____ in play money.

At times you may not be told to which place to round a number. You can round to the greatest place.

Example 3

On Tuesday, the White House had 219 visitors. The following day, there were 694 visitors. About how many people visited the White House over the two days?

Estimate 219 + 694.
The greatest place of each number is the _____ place.
Round to this place.

219 ⟶ ☐
+ 694 ⟶ + ☐
―――   ―――
           ☐

Talk MATH

Look at the problem in Example 3. How could it be rewritten so an exact answer is needed?

So, about _____ people visited the White House over the two days.

Guided Practice

Estimate. Round each addend to the indicated place value.

1. 312 + 27; tens

   ____ + ____ = ____

2. 383 + 122; hundreds

   ____ + ____ = ____

88 Chapter 2 Addition

Name _____

Independent Practice

Estimate. Round each addend to the indicated place value.

3. $34 + $23; tens

    _____ + _____ = _____

4. 636 + 27; tens

    _____ + _____ = _____

5. 687 + 231; hundreds

    _____ + _____ = _____

6. 1,624 + 334; hundreds

    _____ + _____ = _____

7. 1,172 + 1,115; tens

    _____ + _____ = _____

8. $4,412 + $1,204; hundreds

    _____ + _____ = _____

Estimate. Round each addend to its greatest place value.

9. 35 + 42

    _____ + _____ = _____

10. 455 + 229

    _____ + _____ = _____

11. 272 + 593

    _____ + _____ = _____

12. 15 + 39

    _____ + _____ = _____

13. 216 + 536

    _____ + _____ = _____

14. 44 + 29

    _____ + _____ = _____

Estimate. Round each addend to the nearest ten and nearest hundred.

		Nearest Ten	Nearest Hundred
15.	133 + 560		
16.	119 + 239		
17.	89 + 71		

Lesson 5 Estimate Sums 89

Problem Solving

18. Processes &Practices 4 **Model Math** About how many racers were in the Summer Fun Race? Write a number sentence to solve.

Summer Fun Race		
Start Time	Group	Entrants
9:00 A.M.	runners	79
10:00 A.M.	race walkers	51

19. To the nearest hundred, what would be a reasonable estimate for the total attendance at the county fair for the two days? Write a number sentence to show your thinking.

Brain Builders

20. Processes &Practices 1 **Make Sense of Problems** A high school gymnasium can seat 2,136 people. To the nearest ten and hundred, about how many people can sit in the gymnasium? Which estimate is closer? Explain.

21. **Building on the Essential Question** When might I need to estimate a sum?

90 Chapter 2 Addition

Name _____

MY Homework

Lesson 5

Estimate Sums

Homework Helper

Need help? connectED.mcgraw-hill.com

The parking lot in front of the school has 153 parking spaces. The parking lot behind the school has 138 spaces. About how many parking spaces are there in all?

Estimate 153 + 138.

Round, and then add.

One Way Hundreds Place

hundreds	tens	ones
1	5	3	→ 200
+ 1	3	8	→ + 100
			300

To the nearest hundred, the parking lot has about 300 parking spaces.

Another Way Tens Place

hundreds	tens	ones
1	5	3	→ 150
+ 1	3	8	→ + 140
			290

To the nearest ten, the parking lot has about 290 parking spaces.

Both estimates are reasonable.

Practice

Estimate. Round each addend to the indicated place value.

1. 34 + 65; tens

    _____ + _____ = _____

2. 583 + 321; hundreds

    _____ + _____ = _____

3. 591 + 234; hundreds

    _____ + _____ = _____

4. $3,327 + $1,548; tens

    _____ + _____ = _____

5. 2,613 + 3,177; tens

    _____ + _____ = _____

6. $251 + $207; hundreds

    _____ + _____ = _____

7. Estimate. Round each addend to the nearest ten and nearest hundred.

	Nearest Ten	Nearest Hundred
363 + 132		

Problem Solving

8. Three large pizzas cost $36. Two medium pizzas cost $25. To the nearest ten, about how much do all 5 pizzas cost?

Brain Builders

9. **Processes &Practices** **Use Number Sense** Four hundred ninety-one people attended the school play and 422 people attended the band concert. Ryan rounds to the nearest hundred and estimates that 800 people attended these two events all together. Is his estimate correct? Explain.

Vocabulary Check

10. Use each word below to explain how you solved Exercise 9.

    round   estimate

11. **Test Practice** Which of the following can be used to estimate the sum of 380 and 437 to the nearest hundred?

    Ⓐ 300 + 400   Ⓒ 380 + 430

    Ⓑ 400 + 400   Ⓓ 380 + 440

Name

Lesson 6
Hands On
Use Models to Add

ESSENTIAL QUESTION
How can place value help me add larger numbers?

Use a place-value chart and base-ten blocks to model three-digit addition with regrouping. To **regroup** means to rename a number using place value.

Build It

While on a trip, Rosa counted 148 red cars and 153 green cars. How many total cars did Rosa count?

Find 148 + 153.

1. Estimate the sum

   148 → ☐
   + 153 → + ☐
   ─────────
            ☐

hundreds	tens	ones

2. Model 148 + 153.

   Draw your models at the right. Use a ☐ for hundreds, | for tens, and ▫ for ones.

3. Add the ones.
   Regroup 10 ones as 1 ten.

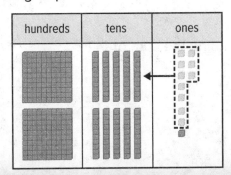

Draw your models.

hundreds	tens	ones

Online Content at connectED.mcgraw-hill.com

Lesson 6 Hands On: Use Models to Add 93

Add the tens and the hundreds.

Regroup 10 tens as _____ hundred. Draw your models below.

hundreds	tens	ones

There are _____ hundreds, _____ tens, and _____ one.

So, 148 + 153 = _____.

Check for Reasonableness

Ask yourself if the answer makes sense. Is your answer **reasonable**?

301 is close to the estimate of 300. It makes sense. The answer is reasonable.

Talk About It

1. Explain how you know when you need to regroup.

_____

_____

2. **Processes &Practices** **Be Precise** Why were the ones and tens regrouped?

_____

_____

3. Tell whether or not you need to regroup when finding the sum of 147 and 214. Explain.

_____

_____

94 Chapter 2 Addition

Name

Practice It

Use models to add. Draw the sum.

4. 259 + 162 = _____

5. 138 + 371 = _____

6. 541 + 169 = _____

7. 261 + 139 = _____

8. 342 + 204 = _____

9. 193 + 154 = _____

Lesson 6 Hands On: Use Models to Add

Apply It

Use base-ten blocks, the table, and the information below to solve each problem.

The Smith family traveled from Chicago, Illinois to Indianapolis, Indiana. Then they traveled from Indianapolis to Memphis, Tennessee.

City to City	Distance
Chicago to Indianapolis	181 miles
Indianapolis to Memphis	464 miles

10. How many miles did the Smith family travel from Chicago, IL, to Memphis, TN? Write a number sentence.

    _____ + _____ = _____ total miles

11. To the nearest hundred, about how many miles did the Smith family travel round trip?

    _____ miles + _____ miles = _____ miles round trip

12. **Processes &Practices 4** **Model Math** The Smith family spent a total of $2,345 on travel expenses and $500 on gasoline. About how much money did they spend all together? Round to the nearest hundred.

    $_____ + $_____ = $_____

Write About It

13. How do I know if I need to regroup when finding a sum?

    _____
    _____
    _____
    _____

96 Chapter 2 Addition

Name ....

Lesson 6

Hands On: Use Models to Add

Homework Helper

Need help? connectED.mcgraw-hill.com

How many total miles did Carrie travel?

City to City	Distance
Buffalo to Harrisburg	186 miles
Harrisburg to Philadelphia	105 miles

1 Estimate the sum.
186 + 105 → 200 + 100 = 300 ← Use your estimate to check for reasonableness later.

2 Model 186 + 105.

3 Regroup 10 ones for 1 ten.

4 Add the tens and the hundreds.

There are 2 hundreds, 9 tens, and 1 ones.

186 + 105 = 291

Carrie traveled 291 miles.

Check

291 is close to the estimate of 300. The answer is reasonable.

Practice

1. Draw place-value blocks to show the sum.

272 + 119 = _____

Lesson 6 My Homework 97

Draw place-value blocks to show the sum.

2. 632 + 354 = _____

3. 216 + 775 = _____

Problem Solving

4. **Processes &Practices** **Model Math** Luisa has 183 pennies. Her dad gives her 128 more. Write a number sentence to show how many pennies Luisa has now.

_____

5. Jonah has read 265 pages. He has 147 more to read. How many pages will he read in all?

_____

Vocabulary Check

Choose the correct word to complete each sentence.

    reasonable regroup sum

6. _____ means to rename a number using place value.

7. The answer to an addition sentence is called the _____.

8. Estimate the exact answer before solving the problem to see if your answer is _____.

Name
....

Lesson 7
Add Three-Digit Numbers

ESSENTIAL QUESTION
How can place value help me add larger numbers?

When you add, you may need to regroup. To **regroup** means to rename a number using place value.

 Math in My World

Example 1

During a weekend trip to Vermont, the Wildlife Club saw **127** wrens and **58** eagles. How many wrens and eagles did the Wildlife Club see?

Find 127 + 58.

hundreds	tens	ones
1	2	7
	5	8

Estimate 127 + 58 ⟶ _____ + _____ = _____

 Add the ones.
7 ones + 8 ones = 15 ones
Regroup 15 ones as 1 ten and 5 ones.

 Add the tens and hundreds.
1 ten + 2 tens + 5 tens = 8 tens
1 hundred + 0 hundreds = 1 hundred

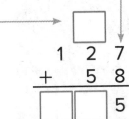

Check for Reasonableness

185 is close to the estimate of _____. It makes sense. The answer is **reasonable.**

127 + 58 = _____

So, the Wildlife Club saw _____ wrens and eagles.

Online Content at connectED.mcgraw-hill.com   Lesson 7 Add Three-Digit Numbers **99**

You can write a number sentence to find the **unknown,** or missing number.

Example 2

One box of butterfly nets costs $175. Another box costs $225. How much do the boxes of nets cost all together?

Write a number sentence to find the unknown.

$175 + $225 = ▪.   ← The unknown is what you are solving for.

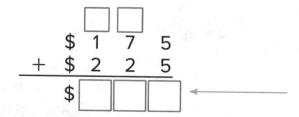

Add the ones.
5 ones + 5 ones = 10 ones
Regroup 10 ones as 1 ten and 0 ones.

Add the tens and hundreds.
1 ten + 7 tens + 2 tens = 10 tens
Regroup 10 tens as 1 hundred and 0 tens.
1 hundred + 1 hundred + 2 hundreds = 4 hundreds

Check for Accuracy

Use the Commutative Property to check your answer. No matter in which order you add, the sum is the same.

$225
+ $175
$400

So, $175 + $225 = _____. The unknown

is _____. The nets cost _____.

Add.

1.
| | 1 | 6 | 4 |
| + | | 1 | 7 |

2.
| | 1 | 5 | 6 |
| + | 2 | 2 | 9 |

Talk MATH

Why is it important to check for reasonableness?

Name _____

Independent Practice

Add. Check for reasonableness.

3.
```
   7 5 9
 +   1 9
```
Estimate: _____

4.
```
   4 4 5
 +   2 6
```
Estimate: _____

5.
```
  $  3 4 5
 +$    9 3
```
Estimate: _____

6.
```
   $427
 + $217
```
Estimate: _____

7.
```
   597
 +  51
```
Estimate: _____

8.
```
   279
 +  19
```
Estimate: _____

Add. Use the Commutative Property of Addition to check for accuracy.

9.
```
   228
 + 149
```

```
   149
 + 228
```

10.
```
   231
 + 596
```

```
   596
 + 231
```

Algebra Add to find the unknown.

11. 43 + 217 = ■

+

The unknown is _____.

12. 607 + 27 = ■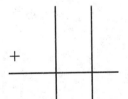

The unknown is _____.

13. $173 + $591 = ■

The unknown is _____.

Lesson 7 Add Three-Digit Numbers **101**

Problem Solving

Write a number sentence with a symbol for the unknown. Then solve.

14. A 10-speed bike is on sale for $199, and a 12-speed racing bike is on sale for $458. How much do the two bikes cost all together?

_____

15. **Processes &Practices 2** **Use Algebra** A newspaper surveyed 475 students about their favorite sport. A magazine surveyed 189 students about their favorite snack. How many students were surveyed by the magazine and newspaper?

_____

Brain Builders

16. **Processes &Practices 2** **Use Number Sense** Use the digits 3, 5, and 7 to make two three-digit numbers with the greatest possible sum. Use each digit one time in each number. Write a number sentence. Explain your reasoning.

_____

_____

17. **Building on the Essential Question** How can I use place value to add three-digit numbers?

_____

_____

_____

_____

Name _____

MY Homework

Lesson 7

Add Three-Digit Numbers

Homework Helper

Need help? connectED.mcgraw-hill.com

A toy store sold 223 robots last year. This year, they sold 198 robots. How many robots did they sell over the two years?

Find 223 + 198.

Estimate   223 + 198 ⟶ 200 + 200 = 400

1 Add the ones.
3 ones + 8 ones = 11 ones
Regroup 11 ones as 1 ten and 1 one.

2 Add the tens and hundreds.
1 ten + 2 tens + 9 tens = 12 tens
Regroup 12 tens as 1 hundred and 2 tens.
1 hundred + 2 hundreds + 1 hundred = 4 hundreds

```
  1 1
  2 2 3
+ 1 9 8
-------
  4 2 1
```

Check for reasonableness

421 is close to the estimate of 400. The answer is reasonable.

So, 223 + 198 = 421.

The store sold 421 toy robots over the two years.

Practice

Add. Check for reasonableness.

1.
```
  1 7 8
+   9 9
-------
```
Estimate: _____

2.
```
  6 9 5
+ 1 4 1
-------
```
Estimate: _____

3.
```
  $ 3 2 7
+ $   5 6
---------
```
Estimate: _____

Lesson 7 My Homework 103

Add. Use the Commutative Property to check for accuracy.

4.
```
  $ 3 5 0
+ $ 4 6 5
─────────
  $
```

```
  $ 4 6 5
+ $ 3 5 0
─────────
  $
```

5.
```
    1 9 6
+   2 8 6
─────────
```

```
    2 8 6
+   1 9 6
─────────
```

Brain Builders

Algebra Add to find the unknown. Explain how you used place value to find the unknown.

6. 661 + 99 = ■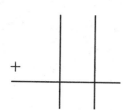

7. $258 + $337 = ■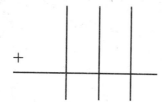

8. $739 + $81 = ■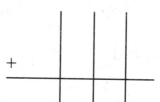

The unknown is ☐.   The unknown is ☐.   The unknown is ☐.

9. **Processes &Practices** **Make a Plan** The principal ordered 215 muffins and 155 bagels. Then he ordered 45 more bagels. How many muffins and bagels were ordered in all?

Vocabulary Check

Choose the correct word to complete each sentence.

   reasonable   regroup   unknown

10. To _____ means to rename a number using place value.

11. A missing number is the _____.

12. If an answer makes sense, it is _____.

13. **Test Practice** Mrs. Lewis bought two statues for her garden. One cost $145 and one cost $262. What was the total cost of the statues?

   Ⓐ $117   Ⓒ $407

   Ⓑ $317   Ⓓ $410

Check My Progress

Vocabulary Check

> addend   addition sentence   estimate   reasonable
>
> regroup   round   sum   unknown

Label each with the correct vocabulary word(s).

1. 139 ⟶ 100
   +273 ⟶ +300
   ———————
   400

   _____

2. 139 + 273 = 412

   412 is close to the estimate of 400.

   The answer is _____.

3. 148 + 153 = 301 ⟵ _____

4. 543 + 281 = ■
   ↑   ↑   ↑

   _____

5. Exchange 10 ones as 1 ten.

hundreds	tens	ones

6. 153 to the nearest ten is 150.
   153 to the nearest hundred is 200.

   _____

Concept Check

Estimate. Round each addend to the indicated place value.

7. 214 + 62; tens

   ____ + ____ = ____

8. 483 + 112; hundreds

   ____ + ____ = ____

Use models to add. Draw the sum.

9. 99 + 209 = _____   10. 316 + 284 = _____   11. 377 + 308 = _____

Add. Check for reasonableness. Explain how you know your answers are reasonable.

12. 34 + 727 =

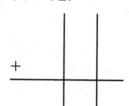

Estimate: _____

13. 528 + 149 =

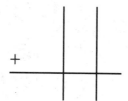

Estimate: _____

14. $193 + $619 =

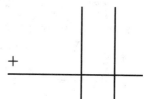

Estimate: _____

Brain Builders

15. Dory's Farm has 468 apple trees and 224 pear trees. To the nearest ten, how many apple and pear trees are there all together? To the nearest hundred, how many apple and pear trees are there all together? Which estimate is better? Explain.

16. There are 369 boys and 421 girls signed up for Adventure Camp this summer. There are 75 counselors. How many people will be at camp?

17. **Test Practice** It will cost $177 a night to stay at a hotel Friday and Saturday. It will cost $159 to stay on Sunday. The Taylors want to stay three nights. How much will it cost to stay three nights?

   Ⓐ $336   Ⓑ $344   Ⓒ $513   Ⓓ $531

Name _____

Lesson 8
Add Four-Digit Numbers

ESSENTIAL QUESTION
How can place value help me add larger numbers?

 Math in My World

4,376 Passengers ALOHA!

Example 1

The largest cruise ship can carry a crew of 1,365. What is the total number of passengers and crew that can travel on this ship?

Find 1,365 + 4,376.

Estimate 1,365 + 4,376 ⟶ 1,400 + 4,400 = _____

 Add the ones.
5 ones + 6 ones = 11 ones
Regroup 11 ones as 1 ten and 1 one.

 Add the tens.
1 ten + 6 tens + 7 tens = 14 tens
Regroup 14 tens as 1 hundred and 4 tens.

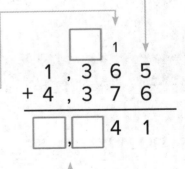

Add the hundreds and thousands.
1 hundred + 3 hundreds + 3 hundreds = 7 hundreds
1 thousand + 4 thousands = 5 thousands

Check for reasonableness.
5,741 is close to the estimate 5,800.

So, 1,365 + 4,376 = _____. The ship can carry _____ people aboard.

Online Content at connectED.mcgraw-hill.com

Lesson 8 Add Four-Digit Numbers 107

Example 2

Last year $3,295 was spent on a skate park. This year $3,999 was spent. How much money was spent over the two years?

Write a number sentence with a symbol for the unknown.

$3,295 + $3,999 = ■

A **bar diagram**, a model used to illustrate a number relationship, may help you organize the information.

| $3,295 | $3,999 |
| last year | this year |

Add.

```
      1   1   1
  $   3,  2   9   5
+ $   3,  9   9   9
  $  ☐   ☐   ☐   ☐
```

⟵ **Check for Accuracy** ⟶

Use the Commutative Property to check your answer. No matter in which order you add, the sum is the same.

```
      1   1   1
  $   3,  9   9   9
+ $   3,  2   9   5
  $  ☐   ☐   ☐   ☐
```

So, $3,295 + $3,999 = _____. The unknown is _____.

Over the two years, _____ was spent on the skate park.

Talk MATH

How could you use the Commutative Property to check that your answer to Exercise 2 is correct?

Guided Practice

Add. Check for reasonableness.

1.
```
    3, 3 4 5
  +    6 5 4
   ☐ ☐ ☐ ☐
```
Estimate:

_____ + _____ = _____

2.
```
  $ 4, 2 3 4
+ $ 1, 7 0 9
  $ ☐ ☐ ☐ ☐
```
Estimate:

_____ + _____ = _____

108 Chapter 2 Addition

Name _____

Independent Practice

Add. Check for reasonableness.

3. 6,499
  +   543
  _____

4. 1,998
  +   300
  _____

5. $2,503
  + $2,899
  _____

6. $8,285
  + $1,456
  _____

7. 2,390
  + 3,490
  _____

8. $5,555
  + $3,555
  _____

Add. Use the Commutative Property of Addition to check for accuracy.

9. 1,734   2,882
  + 2,882   + 1,734
  _____   _____

10. 2,333   5,977
  + 5,977   + 2,333
  _____   _____

Algebra Add to find the unknown.

11. 2,865 + 5,522 = ■

12. 3,075 + 5,640 = ■

13. 1,603 + 3,509 = ■

The unknown is _____. The unknown is _____. The unknown is _____.

Lesson 8 Add Four-Digit Numbers   **109**

Problem Solving

14. In one year, Lou's dad used 1,688 gallons of gas in his car. His mom's car used 1,297 gallons. Use a bar diagram to find the total gallons of gas used. Write a number sentence with a symbol for the unknown. Then solve.

|-------- ■ total gallons --------|
| | |

15. Processes &Practices **3** **Draw a Conclusion** About how many people were surveyed about their favorite summer place? Is the estimate greater than or less than the exact answer? Explain.

Beach   2,311
Amusement Park 2,862

16. Processes &Practices **2** **Use Number Sense** Use the digits 0 through 7 to create two 4-digit numbers whose sum is greater than 9,999. Is there more than one possible answer? Explain.

17. **Building on the Essential Question** Explain how you can check your answer for reasonableness. Apply this strategy to check your answer for Exercise 14.

110 Chapter 2 Addition

Name

MY Homework

Lesson 8

Add Four-Digit Numbers

Homework Helper

Need help? connectED.mcgraw-hill.com

The circus had an attendance of 7,245 people in the bleachers and 1,877 people in the box seats for the Friday night show. How many people attended the circus all together?

Find 7,245 + 1,877.

Estimate 7,200 + 1,900 = 9,100

1. Add the ones.
5 ones + 7 ones = 12 ones
Regroup 12 ones as 1 ten and 2 ones.

2. Add the tens.
1 ten + 4 tens + 7 tens = 12 tens
Regroup 12 tens as 1 hundred and 2 tens.

```
  1 1 1
  7,2 4 5
+ 1,8 7 7
---------
  9,1 2 2
```

3. Add the hundreds and thousands.
1 hundred + 2 hundreds + 8 hundreds = 11 hundreds
Regroup 11 hundreds as 1 thousand and 1 hundred.
1 thousand + 7 thousands + 1 thousand = 9 thousands

Check for reasonableness.
9,122 is close to the estimate of 9,100.
The answer is reasonable.

7,245 + 1,877 = 9,122.
On Friday night, 9,122 people attended the circus.

Practice

Add. Check for reasonableness.

1.
```
   4, 0 9 1
+  2, 2 3 8
```

2.
```
  $ 5, 0 4 5
+ $ 3, 9 9 9
```

Lesson 8 My Homework 111

Add. Check for reasonableness.

3.
```
   2, 0 8 8
+  6, 3 4 6
```

4.
```
   4, 4 6 3
+  4, 8 1 9
```

5.
```
   3, 8 6 6
+  4, 7 2 7
```

6. Use the Commutative Property to check your answer to Exercise 3.

```
   6, 3 4 6
+  2, 0 8 8
```

Brain Builders

Algebra Add to find each unknown.

7. 7,028 + 2,578 = ■
```
   7, 0 2 8
+  2, 5 7 8
```
The unknown is _____.

8. 5,724 + 2,197 = ■
```
   5, 7 2 4
+  2, 1 9 7
```
The unknown is _____.

9. 4,999 + 4,265 = ■
```
   4, 9 9 9
+  4, 2 6 5
```
The unknown is _____.

What is one way to check your answer for reasonableness?

10. **Processes & Practices 2** **Use Algebra** Rachael ran 3,012 meters on Monday and 5,150 meters on Wednesday. How many meters did she run all together? Complete the diagram. Write a number sentence with a symbol for the unknown. Then solve.

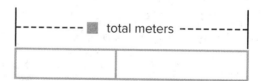

11. **Test Practice** Mr. Shelton worked 1,976 hours one year and 2,080 hours the next year. He also worked 138 hours overtime. What is the total number of hours Mr. Shelton worked in those two years?

Ⓐ 4,084 hours   Ⓒ 4,156 hours

Ⓑ 4,194 hours   Ⓓ 4,056 hours

Name ....

Lesson 9
Problem-Solving Investigation
STRATEGY: Reasonable Answers

ESSENTIAL QUESTION
How can place value help me add larger numbers?

Learn the Strategy

Is it reasonable to say that Wallace's family has traveled about 2,200 miles this year?

1 Understand

What facts do you know?

I know Wallace's family has traveled

_____ miles, _____ miles, and _____ miles.

What do you need to find?

I need to find if the family traveled about _____ miles.

This Year's Travel	
Dad	398 miles
Wallace	737 miles
Brother	1,106 miles

2 Plan

I will estimate the sum. Then I will compare the _____.

3 Solve

Estimate 1,106 ⟶ 1,100
         737 ⟶   700
     + 398 ⟶ + 400

The estimated sum of _____ miles is the same as the estimate in the problem. So, the answer is reasonable.

4 Check

Does your answer make sense? Explain.

Yes. 398 + 737 + 1,106 = 2,241, which is close to the estimated sum of _____.

Online Content at ⌖ connectED.mcgraw-hill.com   Lesson 9 Problem-Solving Investigation **113**

Practice the Strategy

Anson swam 28 laps last week and 24 laps this week. He says he needs to swim the same number of laps each week for two more weeks to swim a total of about 100 laps. Is this a reasonable estimate? Explain.

1 Understand

What facts do you know?

_____

_____

What do you need to find?

_____

_____

2 Plan

_____

_____

3 Solve

4 Check

Does your answer make sense? Explain.

_____

114 Chapter 2 Addition

Name ....

Apply the Strategy

Determine a reasonable answer for each problem.

1. A bank account has $3,701 in it on Monday. On Tuesday, $4,294 is added to the account. Is it reasonable to say that now there is about $7,000 in the account? Explain.

2. Last year, 337 books were sold at the book fair. This year, 217 more books were sold than last year. Is it reasonable to say that more than 500 books were sold this year? Explain.

Brain Builders

3. Juliana estimated that she needs to make 100 favors for the family reunion. Is this a reasonable estimate if 32 family members come on Thursday, 35 family members come on Friday, and 42 family members come on Saturday? Explain.

4. **Processes &Practices** 3 **Justify Conclusions** A goal of 9,000 people in attendance was set for the State fair Friday and Saturday.

State Fair Attendance	
Day	Attendance
Friday	5,653
Saturday	4,059

Is it reasonable to say that the goal was met? Explain.

Lesson 9 Problem-Solving Investigation **115**

Review the Strategies

Use any strategy to solve each problem.
- Use the four-step plan.
- Determine reasonable answers.

5. How much will all of the flowers cost?

Mr. White's Garden Shop		
Flower	Quantity	Cost Each
Daisy	7	$5
Rose	3	$10
Lily	4	$6
Petunia	9	$4
Marigold	9	$3

6. **Processes &Practices** 2 **Use Symbols** The cards show the number of points Sylvia and Meli have in a game.

Sylvia's cards

Meli's cards

How many points do Sylvia and Meli each have? Who has the greater number of points? Use < or >.

7. Val spent $378 at the mall. Her sister spent $291. About how much did the sisters spend together?

116 Chapter 2 Addition

Name

MY Homework

Lesson 9

Problem Solving: Reasonable Answers

Homework Helper

Need help? connectED.mcgraw-hill.com

Is it reasonable to say that about 3,000 students participated in the survey?

Career Survey	
Scientist	2,129 votes
Writer	1,093 votes
Doctor	1,076 votes

1 Understand

What facts do you know?

2,129 students voted for scientist, 1,093 students voted for writer, and 1,076 students voted for doctor.

What do you need to find?

If 3,000 is a reasonable total for all the votes.

2 Plan
Estimate the sum. Then, compare the estimate to 3,000.

3 Solve

```
  2,129  ⟶    2,100
  1,093  ⟶    1,100
+ 1,076  ⟶  + 1,100
             ──────
              4,300
```

The estimated sum of 4,300 is not close to 3,000.
So, 3,000 is not a reasonable total for all the votes.

4 Check
Does your answer make sense? Explain.

Yes; 2,129 + 1,093 + 1,076 is 4,298, which is not close to the suggested total of 3,000. 3,000 is not a reasonable total.

Lesson 9 My Homework 117

Problem Solving

Determine a reasonable answer for each problem.

1. Students have borrowed 632 fiction books and 392 non-fiction books from the school library. Is it reasonable to say the students borrowed about 1,000 books? Explain.

2. Some students from Sydney Elementary have entered a National Poetry contest. There are 19 second graders, 23 third graders, and 9 fourth graders who have entered. Are there at least 50 students entered in the contest? Explain.

Brain Builders

3. **Processes & Practices 2** **Reason** The head of Gracie's favorite dinosaur is 14 feet long, the body is 26 feet long, and the tail is 19 feet long. Is it reasonable to say the entire dinosaur is 60 feet long? Explain.

Use the table to solve Exercises 4 and 5.

Radio Stations	
Country	3,589 listeners
Rock	2,986 listeners
Hip Hop	3,472 listeners

4. How many people listen to country and hip hop? What role did place value play in finding the sum?

5. The goal for the rock and country stations was to have 7,000 listeners. Is it reasonable to say that the stations reached their goal? Explain.

Fluency Practice

Add.

1. 54
   +43

2. 36
   +12

3. 74
   +32

4. 63
   +26

5. 52
   +27

6. 64
   +18

7. 43
   +39

8. 89
   +75

9. 91
   +36

10. 88
    +57

11. 77
    +39

12. 82
    +81

13. 710
    + 33

14. 811
    + 49

15. 99
    +28

16. 800
    + 64

17. 426
    +319

18. 109
    + 72

19. 293
    +310

20. 365
    +364

Name ....

Fluency Practice

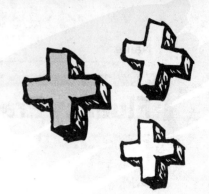

Add.

1. 66
   +34

2. 73
   +19

3. 91
   +51

4. 163
   + 77

5. 25
   +96

6. 641
   +187

7. 123
   +390

   51
   +45

9. 910
   + 48

10. 888
    + 37

11. 771
    + 98

12. 382
    +591

13. 625
    + 63

14. 754
    +111

15. 990
    + 10

16. 329
    + 34

17. 264
    + 91

18. 428
    +326

19. 531
    + 11

20. 508
    +426

Review

Chapter 2
Addition

Vocabulary Check

Choose the correct word(s) below to complete each sentence.

Associative Property	bar diagram	Commutative Property
estimate	Identity Property	mental math
parentheses	pattern	reasonable
regroup	unknown	

1. The _____ of Addition states that you can change the order of the addends and still get the same sum.

2. _____ tell you which operations to group in a number sentence.

3. _____ is a way of doing math in my head.

4. A number close to its actual value is a(n) _____.

5. To use place value to exchange equal amounts when renaming a number is to _____.

6. A _____ is a set of numbers that follow a certain order.

7. The _____ of Addition states when you add zero to any number, the sum is that number.

8. When an answer makes sense, it is _____.

9. The missing number or what you are solving for in a problem is the _____.

My Chapter Review 121

Concept Check

Find each sum. Identify the property of addition. Write *Commutative*, *Associative*, or *Identity*.

10. 8 + 0 = _____

11. 2 + (4 + 3) = _____

(2 + 4) + 3 = _____

12. 5 + 7 = _____

7 + 5 = _____

Write the number.

13. 1 more than 375

14. 1,000 more than 2,184

Make a ten or a hundred to add mentally.

15. 198 + 132

☐ ☐

_____ + _____ = _____

16. 1,274 + 3,599

☐ ☐

_____ + _____ = _____

Estimate. Round each addend to the indicated place value.

17. 725 + 229; tens

_____ + _____ = _____

18. 8,291 + 1,101; hundreds

_____ + _____ = _____

Add. Check for reasonableness.

19.
6	4	3
+ 2	8	2

Estimate:

_____ + _____ = _____

20.
2,	2	0	8
+ 5,	0	9	2

Estimate:

_____ + _____ = _____

122 Chapter 2 Addition

Name _____

Problem Solving

Write a number sentence with a symbol for the unknown. Then solve.

21. A horse weighs 1,723 pounds. A smaller horse weighs 902 pounds. How much do both horses weigh together?

22. A golf ball has 336 dimples. How many dimples would two golf balls have?

Brain Builders

23. Samuel walked one mile before lunch and one mile after lunch. The length of a mile is 1,760 yards. He walked the same distance the next day. In yards, what is the total distance Samuel walked?

24. There are two different mountains in the state of Washington that get record snowfall. One winter season, Mount Ranier received 1,224 inches of snow. Mount Baker received 1,140 inches of snow. What is the total snowfall of these two mountains? Is your answer reasonable? Explain.

25. **Test Practice** Dwight drove 792 miles from Pensacola, Florida, to Key West, Florida, and then back again. He estimated this distance to the nearest hundred miles and the nearest ten miles. What is the difference in the estimates?

    Ⓐ 10 miles
    Ⓑ 1,580 miles
    Ⓒ 20 miles
    Ⓓ 1,600 miles

My Chapter Review 123

Reflect

Chapter 2

Answering the
ESSENTIAL QUESTION

Use what you learned about addition to complete the graphic organizer.

Real-World Problem

Draw a Model

ESSENTIAL QUESTION
How can place value help me add larger numbers?

Properties

Vocabulary

Now reflect on the ESSENTIAL QUESTION Write your answer below.

Performance Task

Football Season

Randy had a great season playing on his school's football team. The table shows the team's scores in each of the first five games.

Game	Point Total
1	35
2	49
3	38
4	27
5	45

Show all your work to receive full credit.

Part A

Randy's mother went to games 1 and 5. His father went to games 2 and 3. Which parent saw the team score more points? Explain.

Part B

Randy's sister went to the two games with the least scoring sum. Without actually adding, suggest a way to find which two games Randy's sister went to. Explain.

Part C

In game 6, the team scored more points than they did in games 1 and 4 combined, but less than 75 points. Name 4 possible amounts the team scored. Label these amounts in the number line shown.

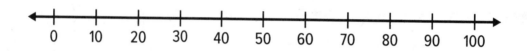

Part D

The scorekeeper made an error. In game 5, the team actually scored 48 points. If you found the sum of all the points scored in the first five games before and after the error was fixed, how would they compare? Explain.

Chapter 3 Subtraction

ESSENTIAL QUESTION
How are the operations of subtraction and addition related?

Activities I Do for Fun

Watch a video!

Name ....

MY Chapter Project

Party Favor Bags

1. As a group, determine the number of guests who will attend a party. You may have as few as 6 and as many as 10.

Number of guests: _____

2. Choose party favors from the list below and write the number of each favor to place in a bag. You may choose up to 10 of each item for a bag.

_____ deck of number cards   _____ package of markers   _____ small notebook

_____ addition facts table   _____ travel game   _____ set of pens

3. Calculate the number of each party favor that will be needed to fill all the bags. For example, if you are making 6 bags and each bag will have 2 travel games, then you will need 12 travel games in all.
$2 + 2 + 2 + 2 + 2 + 2 = 12$

_____   _____   _____

_____   _____   _____

4. Three of the guests cannot attend. Use subtraction to determine the number of bags and the number of party favors that are needed now. Use the space below to show how you found the difference.

Name _____

Am I Ready?

Subtract.

1. 9
   − 4

2. 12
   − 7

3. 15
   − 8

4. 11
   − 6

5. 13 − 7 = ____

6. 10 − 6 = ____

7. 9 − 6 = ____

8. 16 − 8 = ____

Use the base-ten blocks to find each difference.

9.

24 − 11 = ____

10.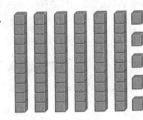

65 − 24 = ____

Round to the nearest ten.

11. 454 ____

12. 689 ____

13. 712 ____

Round to the nearest hundred.

14. 377 ____

15. 409 ____

16. 1,335 ____

Shade the boxes to show the problems you answered correctly.

How Did I Do? | 1 | 2 | 3 | 4 | 5 | 6 | 7 | 8 | 9 | 10 | 11 | 12 | 13 | 14 | 15 | 16 |

Online Content at connectED.mcgraw-hill.com

Name _____

MY Math Words

Review Vocabulary

add	addend	difference
equal	equals sign (=)	estimate
minus sign (−)	plus sign (+)	subtract
subtraction sentence	sum	

Making Connections

Use the review vocabulary to complete the Venn diagram.

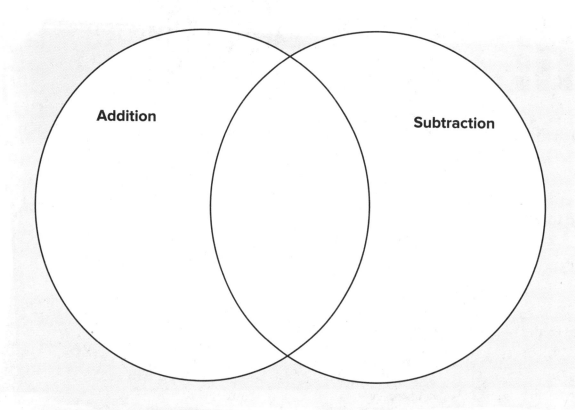

128 Chapter 3 Subtraction

MY Vocabulary Cards

Lesson 3-4

inverse operations

$5 + 3 = 8$
$8 - 3 = 5$

Lesson 3-4

regroup

→ 130

Ideas for Use

- Write a tally mark on each card every time you read the word in this chapter or use it in your writing. Challenge yourself to use at least 10 tally marks for each word card.

- Use the blank cards to write your own vocabulary cards.

To use place value to exchange equal amounts when renaming a number.

The prefix *re-* can mean "again." Write two other math words that use the prefix *re-*.

Operations that undo each other.

Inverse can mean "opposite." How can this help you remember the meaning of *inverse operations*?

MY Foldable

FOLDABLES Follow the steps on the back to make your Foldable.

Regroup with 4-Digit Numbers

Subtracting Across Zeros

Estimating Difference

Regroup with 3-Digit Numbers

There are 5,395 men and women in a race.
There are 2,697 men.
How many runners are women?

5,395
− 2,697
―――――

On Saturday, there were 1,000 balloons at a hot air balloon festival. On Sunday, there were 752 balloons.
How many more balloons were there on Saturday than on Sunday?

1,000
− 752
―――――

The school's lacrosse game was attended by 244 students and 117 parents.
To the nearest ten, about how many more students attended the game?

244 → 240
117 → −120
―――――

There were 381 students who voted for a field trip to the aquarium and 125 students who voted for a field trip to the museum.
How many more students voted for the aquarium?

381
− 125
―――――

Name _____

Lesson 1
Subtract Mentally

ESSENTIAL QUESTION
How are the operations of subtraction and addition related?

To subtract mentally, break up the smaller number into parts. Then subtract in parts.

 Math in My World

Example 1

It was a sunny, warm 86°F day for outdoor games. What was the temperature when it dropped 17°F?

Find 86 − 17.

 Break apart 17.   86 − 17 (17 = 16 + 1)

86 and 16 end in the same digit.

 Subtract 16.   86 − 16 = _____

 Subtract 1.   70 − 1 = _____

So, 86 − 17 = 69. The temperature was 69°F by the end of the day.

You can also use subtraction rules to subtract mentally.

Subtraction Rules	
Subtracting a number from itself equals zero.	367 − 367 = 0
Subtracting zero from a number equals itself.	545 − 0 = 545

Example 2

Find 417 − 417.

Subtracting a number from itself equals _____.

So, 417 − 417 = _____.

Online Content at connectED.mcgraw-hill.com

Lesson 1   133

You can mentally subtract a number that ends in 9 or 99.

Example 3

Find 140 − 129.

Make a ten.
129 is close to 130.

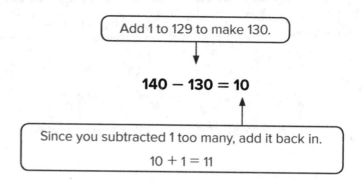

So, 140 − 129 = 11.

Example 4

Find 223 − 99.

Make a hundred.
99 is close to 100.

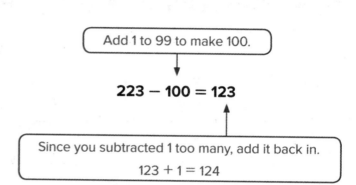

So, 223 − 99 = _____.

Guided Practice

1. Mentally subtract 34 − 18 by breaking apart the smaller number.

   18 = 14 + _____

   34 − 14 = _____

   _____ − 4 = _____

   So, 34 − 18 = _____.

Talk MATH

What mental subtraction strategy could you use to find 234 − 29?

2. Mentally subtract 94 − 59 by making a ten.

   94 − 60 = _____

   _____ + 1 = _____

   So, 94 − 59 = _____.

134 **Chapter 3** Subtraction

Name _____

Independent Practice

Subtract mentally by breaking apart the smaller number.

3. 792 − 94 = _____

4. 885 − 52 = _____

5. 831 − 321 = _____

6. 725 − 717 = _____

Make a 10 or 100 to subtract mentally.

7. 87 − 69 = _____

8. 745 − 239 = _____

9. 652 − 599 = _____

10. 384 − 199 = _____

Write the number sentences under their subtraction rule.

11. When you subtract a number from itself, you get 0.

12. When you subtract 0 from a number, you get that number.

_____

_____

_____

_____

1,937 − 1,937 = 0

9,999 − 0 = 9,999

4,274 − 0 = 4,274

491 − 491 = 0

Lesson 1 Subtract Mentally 135

Problem Solving

Use a mental subtraction strategy to solve.

13. How much could Maurice have saved on his shoes if he waited until today to buy them?

14. **Processes &Practices 1** **Keep Trying** Blair started the morning with $75. Every activity she did today cost money. Find how much money she had at the end of the day.

Brain Builders

15. **Processes &Practices 5** **Use Mental Math** Write each difference as you subtract mentally. Explain why some numbers are easier than others to subtract mentally.

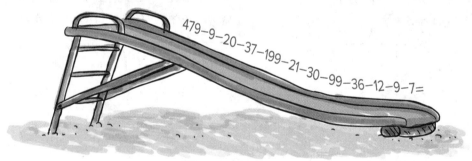

16. **Building on the Essential Question** How can I subtract mentally?

136 Chapter 3 Subtraction

Name _____

MY Homework

Lesson 1

Subtract Mentally

Homework Helper

Need help? connectED.mcgraw-hill.com

Mia wants a guitar that costs **$96**. So far, she has saved up **$48**. How much more money does she need to save in order to buy the guitar?

Find $96 − $48. Subtract in parts.

1. Break apart 48. $96 − $48 ($48 = **$46** + **$2**)

2. Subtract one part. $96 − **$46** = $50

3. Subtract the other part. $50 − **$2** = $48

So, Mia needs $48 more.

Practice

Subtract mentally by breaking apart the smaller number.

1. 82 − 47 = _____

2. 165 − 26 = _____

3. 387 − 308 = _____

4. 674 − 426 = _____

Make a 10 or 100 to subtract mentally.

5. 76 − 59 = _____

6. 120 − 39 = _____

7. 554 − 199 = _____

8. 453 − 19 = _____

Lesson 1 My Homework 137

Problem Solving

Use any mental math strategy to solve.

9. **Processes & Practices 5** **Use Mental Math** The red team has 522 fans in the bleachers. The blue team has 425 fans in the bleachers. How many fewer fans does the blue team have?

Brain Builders

10. **Processes & Practices 2** **Use Number Sense** There are 172 houses in Kyle's neighborhood. He delivers papers to 99 houses. To how many houses does Kyle *not* deliver papers? How can you check your answer?

Vocabulary Check

Draw a line to match the word to its definition or meaning.

11. subtract • The answer to a subtraction problem.

12. difference

• An operation that tells the difference between two numbers.

13. **Test Practice** Avery has a digital camera with a memory card that can hold 284 pictures. Avery has taken 159 pictures of dogs and 32 pictures of cats. How many pictures can the memory card still hold?

Ⓐ 93 pictures

Ⓑ 95 pictures

Ⓒ 125 pictures

Ⓓ 475 pictures

138 **Need more practice?** Download Extra Practice at connectED.mcgraw-hill.com

Name _____

Lesson 2
Estimate Differences

ESSENTIAL QUESTION
How are the operations of subtraction and addition related?

Sometimes you want to find an estimate instead of an exact answer. You can estimate differences to the nearest hundred or nearest ten.

  Math in My World

Example 1

China has more than 3,728 miles of high-speed train tracks. Japan has more than 1,518 miles of high-speed tracks. About how many more miles of high-speed tracks does China have?

Round each number in 3,728 − 1,518.

Then find the estimated difference.

To the Nearest Hundred

	thousands	hundreds	tens	ones
3,728 →	3	0	0	0
−1,518 → −	1	0	0	0

To the Nearest Ten

	thousands	hundreds	tens	ones
	3	7	0	0
−	1	5	0	0

To the nearest hundred, there are about _____ more miles of high-speed tracks in China.

To the nearest ten, there are about _____ more miles of high-speed tracks in China.

Each estimate is reasonable. There can be more than one reasonable estimate when solving a problem.

Online Content at connectED.mcgraw-hill.com

You are not always told to which place to round a number.
You can round to the greatest place value.

Example 2

A museum has 237 works of art on its first floor. The second floor has 349 works of art. About how many more works of art are found on the second floor?

Estimate 349 − 237.

The greatest place-value position of 349 and 237 is the

_____ place.

Round to the hundreds place. Then subtract.

$$\begin{array}{r} 349 \\ -237 \end{array} \longrightarrow \begin{array}{r} 300 \\ -200 \end{array}$$ ← Subtract mentally.

So, there are about _____ more paintings on the second floor.

Guided Practice

1. Estimate 488 − 351. Round each number to the tens place.

hundreds	tens	ones

Talk MATH

4,749 was rounded to 4,750. Was 4,749 rounded to the nearest ten or hundred? Explain.

2. Estimate 542 − 225. Round each number to the greatest place value.

hundreds	tens	ones

Name _____

Independent Practice

Estimate. Round each number to the indicated place value.

3. 986 − 664; tens

   _____ − _____ = _____

4. 550 − 244; tens

   _____ − _____ = _____

5. 1,836 − 1,648; hundreds

   _____ − _____ = _____

6. 7,621 − 2,000; hundreds

   _____ − _____ = _____

Estimate. Round each number to the greatest place value.

7. 937 − 338

   _____ − _____ = _____

8. 51 − 24

   _____ − _____ = _____

9. 716 − 207

   _____ − _____ = _____

10. 885 − 474

    _____ − _____ = _____

Estimate. Round each number to the nearest ten and nearest hundred.

		Nearest 10	Nearest 100
11.	632 −313		
12.	877 −770		
13.	584 −341		

Lesson 2 Estimate Differences 141

Problem Solving

14. **Processes &Practices 6** **Be Precise** A theater compared weekend ticket sales. Estimate to the nearest hundred to find in which month fewer tickets were sold. Explain.

Theater Ticket Sales		
Day	January	February
Saturday	3,924	2,945
Sunday	2,789	1,754

15. **Processes &Practices 1** **Make Sense of Problems** Gina's brother is going to camp in the summer. What is the estimated difference in the total cost between attending Camp A and Camp B? Round to the nearest ten.

$ _____

Camp Costs		
Camp	Fee	Other Expenses
Camp A	$1,192	$805
Camp B	$1,055	$979

16. **Building on the Essential Question** How do I know to which place-value position to round a number?

142 Chapter 3 Subtraction

Name _____

MY Homework

Lesson 2

Estimate Differences

Homework Helper

Need help? connectED.mcgraw-hill.com

Noah is buying one of two scooters. About how much money would Noah save if he bought the less expensive scooter? Round to the nearest ten and nearest hundred.

One Way Round to the nearest hundred.

To the nearest hundred, Noah would save $200.

$$\begin{array}{r}\$1,463 \\ -\$1,322\end{array} \longrightarrow \begin{array}{r}\$1,500 \\ -\$1,300 \\ \hline \$\ \ 200\end{array}$$

Another Way Round to the nearest ten.

To the nearest ten, Noah would save $140.

$$\begin{array}{r}\$1,463 \\ -\$1,322\end{array} \longrightarrow \begin{array}{r}\$1,460 \\ -\$1,320 \\ \hline \$\ \ 140\end{array}$$

Each estimate is reasonable. There can be more than one reasonable estimate when solving a problem.

Practice

Estimate. Round each number to the indicated place value.

1. 816 − 708; tens

_____ − _____ = _____

2. 466 − 152; hundreds

_____ − _____ = _____

3. 537 − 288; hundreds

_____ − _____ = _____

4. 9,531 − 1,428; tens

_____ − _____ = _____

Estimate. Round each number to the nearest ten and nearest hundred.

	Nearest 10	Nearest 100
5. 3,677 − 2,232		
6. 573 − 441		
7. 1,885 − 483		

Brain Builders

8. Shannon's scout troop sold 2,357 boxes of cookies. They started with 3,600 boxes. To the nearest hundred, about how many boxes of cookies do they have left to sell? Write a number model to support your answer.

9. **Processes & Practices 2** Use Number Sense Clearview Stadium sold 8,371 bags of peanuts last weekend. Capital Stadium sold 4,309 bags of peanuts last weekend. Which stadium sold more bags of peanuts? To the nearest ten, about how many more bags were sold?

10. **Test Practice** Which shows the estimated difference for 8,859 − 3,591 to the nearest hundred?

    Ⓐ 5,000
    Ⓒ 5,300
    Ⓑ 5,268
    Ⓓ 5,400

Name ....

Lesson 3
Problem-Solving Investigation
STRATEGY: Estimate or Exact Answer

ESSENTIAL QUESTION
How are the operations of subtraction and addition related?

Learn the Strategy

To celebrate Arbor Day, Brita's school district planted trees. The high school students planted 1,536 trees. The elementary students planted 1,380 trees. About how many more trees did the high school students plant?

Understand
What facts do you know?

The high school students planted _____ trees.

The elementary students planted _____ trees.

What do you need to find?

about how many more trees the _____ students planted

Plan
An exact answer is not needed. I will estimate.

Solve

Round each number.   Then, subtract.

1,536 → 1,500   ← Round each to the   1,500
1,380 → 1,400   ← nearest hundred.   − 1,400
                                            100

About 100 more trees were planted by the high school students.

Check
Does your answer make sense? Explain.

_____

Online Content at connectED.mcgraw-hill.com   Lesson 3 Problem-Solving Investigation   **145**

Practice the Strategy

Students in grades 2 and 3 wrote 61 stories to celebrate author day. The second graders wrote 26 stories. How many stories did the third graders write?

1 Understand

What facts do you know?

_____

_____

What do you need to find?

_____

2 Plan

_____

_____

3 Solve

4 Check

Does your answer make sense? Explain.

_____

Name
....

Apply the Strategy

Determine whether each problem requires an estimate or an exact answer. Then solve.

1. **Processes & Practices 4 Model Math** Michelann cut two lengths of rope. One was 32 inches long. The other was 49 inches long. Will she have enough rope for a project that needs 76 inches of rope? Explain.

   _____

   _____

2. **Processes & Practices 2 Use Number Sense** The number 7 septillion has 24 zeros after it. The number 7 octillion has 27 zeros after it. How many zeros is that all together? What is the difference between the number of zeros in 7 septillion and 7 thousand?

   _____

   _____

Brain Builders

3. **Processes & Practices 3 Justify Conclusions** Three buses can carry a total of 180 students. Is there enough room on the 3 buses for 95 boys and 92 girls? Explain.

   _____

   _____

4. Mrs. Carpenter received a bill for car repairs. About how much was the oil change if the other repairs were $102?

   _____

   _____

Car Repair: $134.00

Lesson 3 Problem-Solving Investigation 147

Review the Strategies

Use any strategy to solve each problem.
- Determine reasonable answers.
- Determine an estimate or exact answer.
- Use the four-step plan.

5. Some children took part in a penny hunt. About how many more pennies did Pat find than either of his two friends?

Penny Hunt	
Cynthia	133
Pat	182
Garcia	125

6. **Processes & Practices 1** **Check for Reasonableness** A bank account has $320 in it on Monday. On Tuesday, $629 is added to the account. On Wednesday, $630 is taken out of the account. Is it reasonable to say that there is about $100 in the account after Wednesday? Explain.

_____

_____

7. When the school first opened, its library had 213 books. Today it has more than 650 books. About how many books has the school bought since it first opened? Explain.

_____

_____

8. The Bonilla family spent $1,679 on their vacation. The Turner family spent $983. About how much less did the Turner family spend? Explain.

_____

_____

148 Chapter 3 Subtraction

MY Homework

Lesson 3

Problem Solving: Estimate or Exact Answer

Homework Helper

Need help? connectED.mcgraw-hill.com

What is the difference in the total number of ribbons the girls earned?

Swim Team Ribbons		
Swimmer	Last Year	This Year
Mackenzie	26	31
Zoe	19	33

1 Understand

What facts do you know?
The number of ribbons each girl earned each year.

What do you need to find?
The difference in the total number of ribbons between the girls.

2 Plan

An exact answer is needed. I will add to find the total.
Then I will subtract to find the difference.

3 Solve

Mackenzie   **Zoe**

```
   26           19
 + 31         + 33
 ----         ----
   57           52
```

$57 - 52 = 5$

The difference between the total number of ribbons for each girl is 5.

4 Check

Does your answer make sense? Explain.
Yes; Use the inverse operation of addition.
$5 + 52 = 57$ shows the answer is correct.

Problem Solving

Determine whether each problem requires an estimate or an exact answer. Then solve.

1. Gavin saved $53 in August and $15 in September. Is it reasonable to say he needs $50 more to pay for a karate class that costs $100? Explain.

2. There were 4,569 fans at a soccer game. The visiting team had 1,604 fans cheering. About how many fans cheered for the home team? Explain.

Brain Builders

3. **Processes & Practices 1** **Make Sense of Problems** Luke buys 2 small pumpkins and 1 large pumpkin. How much does Luke spend?

   Pumpkin Sale
   Small Pumpkins $4
   Large Pumpkins $7

1,337 pounds

4. The brown bear at a local zoo weighs 1,578 pounds. The black bear weighs 643 pounds. About how much more do the brown bear and black bear together weigh than the polar bear? Explain.

150 **Need more practice?** Download Extra Practice at connectED.mcgraw-hill.com

Check My Progress

Vocabulary Check

1. Match each word with its definition. Finish drawing the puzzle piece of each word so that it fits the puzzle piece of its definition.

   subtract — a number close to the exact value

   difference — the answer to a subtraction problem

   estimate — to take some or all away

Concept Check

2. Mentally find the difference between the numbers that are connected. Write the difference in the box below the arrow. The last number is given so that you can check your work.

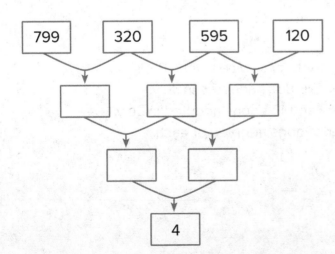

Problem Solving

3. A tent costs $499. It is on sale for $75 off of the original price. Mentally subtract to find the sale price.

4. The table shows how much money was made at a movie theater last week. About how much more was earned on Friday than on Sunday?

Movie Sales	
Day	Amount
Friday	$432
Saturday	$721
Sunday	$184

Determine if an estimate or an exact answer is needed. Then solve.

Brain Builders

5. This year, the third grade raised $379 for a dog shelter. Last year, they raised $232. Mentally subtract to find how much more money was raised this year than last year. Explain your process to a friend.

6. **Test Practice** Two shelters offer dog adoptions twice a year. Ruffs had 123 dogs adopted in January and 196 dogs adopted in July. Barks had 78 dogs adopted in January and 134 dogs adopted in July. Estimate the difference in the number of dogs adopted at each shelter that year.

Ⓐ 100   Ⓒ 200
Ⓑ 107   Ⓓ 531

152 **Chapter 3** Subtraction

Name _____

Lesson 4
Hands On
Subtract with Regrouping

ESSENTIAL QUESTION
How are the operations of subtraction and addition related?

Sometimes you need to **regroup** when subtracting. You will use place value to exchange equal amounts to rename a number.

Build It

Find 244 − 137 = ■. ← unknown

Estimate 244 − 137 ⟶ 200 − 100 = 100

1 Model 244.
Use base-ten blocks. Draw your model.

2 Subtract ones.
You cannot subtract 7 ones from 4 ones.
Regroup 1 ten as 10 ones.
There are now 14 ones.
Subtract 7 ones.
14 ones − 7 ones = ☐ ones.

Online Content at connectED.mcgraw-hill.com

Lesson 4   153

3 Subtract 3 tens.

3 tens − 3 tens = _____ tens

4 Subtract hundreds.

2 hundreds − 1 hundred =

_____ hundred

Draw the result.

So, 244 − 137 = _____.

The unknown is _____.

Check

Addition and subtraction are **inverse operations** because they undo each other.

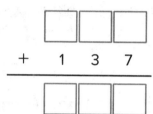

Talk About It

1. In Step 2, why did you regroup 1 ten as 10 ones?

   _____

2. What did you notice about the tens in Step 3 when you subtracted them?

   _____

3. **Processes &Practices** **2** **Stop and Reflect** Suppose, after subtracting the ones, there were not enough tens from which to subtract. What do you think you may have to do?

   _____

4. Why can you use addition to check your answer to a subtraction problem?

   _____

Name

Practice It

Use models to subtract. Draw the difference.

5. 181 − 63 = _____

6. 322 − 118 = _____

7. 342 − 119 = _____

8. 212 − 103 = _____

Use models to subtract. Draw the difference. Use addition to check.

9. 341 − 19 = _____

10. 553 − 128 = _____

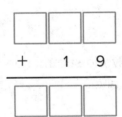

11. 338 − 175 = _____

12. 632 − 313 = _____

Lesson 4 Hands On: Subtract with Regrouping 155

Apply It

Use base-ten blocks to solve.

13. **Processes &Practices 5** **Use Math Tools** The Bank of America in Charlotte, North Carolina, is 871 feet tall. The One Liberty Place building in Philadelphia, Pennsylvania, is 945 feet tall. Write a number sentence to find how much taller the One Liberty Place building is. Then write an addition number sentence to check.

_____ − _____ = _____

The One Liberty Place building is _____ feet taller than the Bank of America building.

_____ + _____ = _____

14. The Storrow family visited the aquarium and saw 483 saltwater fish and 358 freshwater fish. Later that month, workers moved 139 fish to a new aquarium. How many fish remained? Explain how you got your answer.

15. **Processes &Practices 4** **Model Math** Write a real-world subtraction word problem in which regrouping is needed to solve.

Write About It

16. What does it mean to regroup?

156 Chapter 3 Subtraction

Name _____

MY Homework

Lesson 4

Hands On: Subtract with Regrouping

Homework Helper

Need help? connectED.mcgraw-hill.com

Mr. Skylar made 432 birdhouses for the craft fair. Of those, 315 were sold. How many birdhouses are left?

Find 432 − 315.

Estimate 432 − 315 ⟶ 430 − 320 = 110

1 Model 432.

2 Subtract the ones.
Regroup 1 ten as 10 ones.
12 ones − 5 ones = 7 ones

3 Subtract the tens and hundreds.
2 tens − 1 ten = 1 ten
4 hundreds − 3 hundreds = 1 hundred

So, Mr. Skylar has 117 birdhouses left.

Check
Use addition to check subtraction.

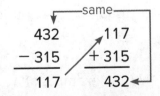

Lesson 4 My Homework 157

Practice

Use models to subtract. Draw the difference.

1. 552 − 361 = _____

2. 636 − 324 = _____

Use models to subtract. Draw the difference. Use addition to check.

3. 486 − 318 = _____

4. 270 − 131 = _____

+ _____

+ _____

Problem Solving

Processes & Practices Use Math Tools Use models to subtract.

5. Nola had $507. She spent $255 on a new bicycle. How much money does Nola have left?

_____

6. Evan set out 643 items for a garage sale. He sold 456 items. How many items does Evan have left?

_____

Vocabulary Check

Choose the correct word(s) to complete each sentence.

inverse operation   regroup   subtraction sentence

7. Addition is the _____ of subtraction.

8. 395 − 278 = 117 is an example of a _____.

9. Sometimes I need to _____, or use place value to rename a number, in order to subtract.

158 Chapter 3 Subtraction

Name _____

Lesson 5
Subtract Three-Digit Numbers

ESSENTIAL QUESTION
How are the operations of subtraction and addition related?

Math in My World

Example 1

How many more sheets of craft paper does Will have than Liseta?

Find the unknown. 265 − 79 = ■

Craft Paper	
Name	Sheets
Liseta	79
Will	265
Alano	128

Estimate   265 → 300
          − 79 → −100
                 ☐

1. Subtract ones.

Regroup 1 ten as 10 ones.

 ← 5 ones + 10 ones = 15 ones

   2 6 **5**
−   7 **9**
   ☐ ← Subtract.

2. Subtract tens and hundreds.

Regroup 1 hundred as 10 tens.

 ← 5 tens + 10 tens = 15 tens

      5 15
   2 **6** 5
−   7   9
   ☐ ☐ 6
     ↘ ↗
     Subtract.

Check

     ←——same——→
   2 6 5   1 8 6
 −   7 9   +   7 9
   ☐   ☐

Addition shows the subtraction answer is correct.

_____ is close to the estimate _____.
Estimation shows the answer is reasonable.

So, 265 − 79 = _____. Will has _____ more sheets of craft paper.

Example 2

Denzel wants to buy a remote control airplane for $125. He has $354. How much money will he have left?

Find the unknown. $354 − $125 = ■. ← unknown

Estimate $354 − $125 ⟶ $350 − $130 = _____

```
            4  14
    $   3  5̸  4̸
  − $   1  2   5
  ─────────────────
    $  □  □  □
```

☐ + 5 = 14
☐ + 2 = 4
☐ + 1 = 3

Helpful Hint: Use addition to help you subtract by thinking of a related fact.

Check

```
            same
   $ 3 5 4      $ 2 2 9
  −$ 1 2 5     +$ 1 2 5
   □□□          □□□
```

Addition shows the answer is correct.

_____ is close to the estimate _____. The answer is reasonable.

So, $354 − $125 = _____. Denzel will have _____ left.

Guided Practice

Subtract. Use addition to check your answer.

1.
```
       □□
   $ 7  6  4
  −$ 1  3  8
   $ □  □  □
```
Check:
```
      $
   +  $
   ─────
      $
```

2.
```
       □
      □□□
   $ 6  1  4
  −$ 4  5  7
   $ □  □  □
```
Check:
```
      $
   +  $
   ─────
      $
```

Talk MATH

Why do you need to rename the tens place twice in Exercise 2?

160 **Chapter 3** Subtraction

Name _____

Independent Practice

Subtract. Use addition to check your answer.

3.
```
   $ | 6 | 8 | 7
 - $ | 3 | 5 | 3
```
Check:

+ _____

4.
```
     | 1 | 7 | 7
 -   |   | 9 | 4
```
Check:

+ _____

5.
```
   $ | 8 | 4 | 3
 - $ | 1 | 8 | 7
```
Check:

+ _____

Algebra Subtract to find the unknown.

6. $769 − $359 = ■

7. 267 − 178 = ■

8. 492 − 383 = ■

The unknown is $ _____.

The unknown is _____.

The unknown is _____.

Algebra Use addition to find each unknown.

9.
```
     | 6 | 1 | ■
 -   | 4 | 1 | 7
     | ▲ | 0 | 2
```
■ = _____

▲ = _____

10.
```
     | ■ | 9 | 9
 -   | 1 | ▲ | 0
     | 2 | 1 | 9
```
■ = _____

▲ = _____

11.
```
     | 7 | 9 | 8
 -   | ■ | 9 | 7
     | 4 | ▲ | 1
```
■ = _____

▲ = _____

Lesson 5 Subtract Three-Digit Numbers **161**

Problem Solving

Glenwood Elementary students were asked to vote for their choice of a field trip destination. The table shows the results.

Field Trip Choices	
Field Trip	Votes
Aquarium	233
Museum	105
Lighthouse	269
Science Center	298

12. How many more students voted for the lighthouse than for the aquarium? Write a number sentence to solve. Then check with an addition sentence.

    _____ − _____ = _____

    _____ + _____ = _____

13. **Processes &Practices** **1** **Check for Reasonableness** How many more students voted for the science center than for the lighthouse? Write a number sentence to solve. Then check with an addition sentence.

    _____ − _____ = _____

    _____ + _____ = _____

Brain Builders

14. **Processes &Practices** **3** **Find the Error** When Federico subtracted 308 from 785, he got 477. To check his answer, he added 308 and 785. What did he do wrong?

15. **Building on the Essential Question** Write a subtraction problem. Explain why you can use addition to check your answer to a subtraction problem.

Name ....

MY Homework

Lesson 5

Subtract Three-Digit Numbers

Homework Helper

Need help? connectED.mcgraw-hill.com

Chloe jumped rope 631 times without stopping. Addison jumped 444 times. How many more jumps did Chloe make?

Find 631 − 444.

1 Subtract ones.

$$\begin{array}{r} \overset{2\,11}{6\cancel{3}\cancel{1}} \\ -\ 444 \\ \hline 7 \end{array}$$

Regroup 1 ten as 10 ones.
10 ones + 1 one = 11 ones
11 ones − 4 ones = 7 ones

2 Subtract tens and hundreds.

$$\begin{array}{r} \overset{12}{5\cancel{2}11} \\ \cancel{6}\cancel{3}\cancel{1} \\ -\ 444 \\ \hline 187 \end{array}$$

Regroup 1 hundred as 10 tens.
10 tens + 2 tens = 12 tens
12 tens − 4 tens = 8 tens
5 hundreds − 4 hundreds = 1 hundred

Check

$$\begin{array}{r} 631 \\ -\ 444 \\ \hline 187 \end{array} \quad \begin{array}{r} 187 \\ +\ 444 \\ \hline 631 \end{array}$$

— same —

Addition shows the subtraction answer is correct.

Since, 631 − 444 = 187, Chloe made 187 more jumps.

Practice

Subtract. Use addition to check your answer.

1.
$$\begin{array}{r} \$\ 5\ 1\ 8 \\ -\ \$\ 3\ 1\ 9 \\ \hline \$\ \square\ \square\ \square \end{array}$$
Check:

2.
$$\begin{array}{r} 6\ 4\ 1 \\ -\ 2\ 2\ 9 \\ \hline \square\ \square\ \square \end{array}$$
Check:

Lesson 5 My Homework 163

Subtract. Use addition to check your answer.

3.
	$	7	6	4
−	$	3	5	3
	$			

Check:

4.
		5	4	2
−		2	6	5

Check:

Algebra Subtract to find the unknown.

5. 599 − 284 = ■

6. 436 − 377 = ■

7. 514 − 175 = ■

The unknown is _____. The unknown is _____. The unknown is _____.

Brain Builders

Write a number sentence to solve.

8. **Processes &Practices 4** **Model Math** Tanisha bought a pack of 225 sheets of paper for her homework. The first week she used 14 sheets of paper. After two weeks, she had 198 sheets of paper left. How many sheets of paper did she use the second week? Show your thinking.

9. The school library would like to raise $915 for new books. In September they raised $280. In October they raised $195. How much more does the library need to raise to reach its goal?

10. **Test Practice** A crate has 272 red and green apples. How many green apples are there?

   Ⓐ 149 green apples   Ⓒ 159 green apples

   Ⓑ 150 green apples   Ⓓ 395 green apples

123 red apples   ? green apples

164 Need more practice? Download Extra Practice at connectED.mcgraw-hill.com

Name ....

Lesson 6
Subtract Four-Digit Numbers

ESSENTIAL QUESTION
How are the operations of subtraction and addition related?

Math in My World

Example 1

What is the difference in height between Ribbon Falls and Kalambo Falls?

NAME	HEIGHT (ft)
Ribbon	1,612
Angel	3,212
Yosemite	2,425
Kalambo	726

Find the unknown. 1,612 − 726 = ▪

Estimate 1,612 − 726 ⟶ 1,600 − _____ = 900

1. Subtract the ones.

Regroup 1 ten as 10 ones.
2 ones + 10 ones = 12 ones

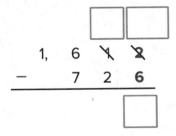

2. Subtract the tens.

Regroup 1 hundred as 10 tens.
0 tens + 10 tens = 10 tens

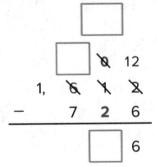

3. Subtract the hundreds and thousands.

Regroup 1 thousand as 10 hundreds.

Check

886 is close to the estimate 900. Estimation shows the answer is reasonable.

So, 1,612 − 726 = _____.

Ribbon Falls is _____ feet taller than Kalambo Falls.

Example 2

Bike Route A is 1,579 miles. Bike Route B is 3,559 miles. How much longer is Bike Route B?

Find the unknown. 3,559 − 1,579 = ?

A ? is a symbol that can be used for the unknown.

Estimate

```
  3, 5 5 9   →    3, 6 0 0
- 1, 5 7 9   →  − 1, 6 0 0
```

 Subtract the ones and tens.

Subtract the ones.
Regroup 1 hundred as 10 tens.
5 tens + 10 tens = 15 tens
Subtract the tens.

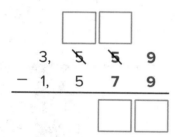

Subtract the hundreds and thousands.

Regroup 1 thousand as 10 hundreds.
10 hundreds + 4 hundreds = 14 hundreds
Subtract the hundreds.
Subtract the thousands.

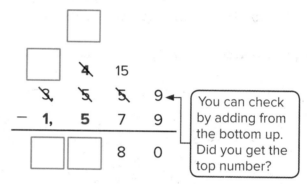

You can check by adding from the bottom up. Did you get the top number?

Bike Route B is _____ miles longer. The unknown is _____.

Check 1,980 is close to the estimate of 2,000. The answer is reasonable.

Guided Practice

1. Subtract. Use addition to check your answer.

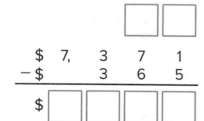

Check:

Talk MATH

Explain the steps to find 8,422 − 5,995.

166 **Chapter 3** Subtraction

Name

Independent Practice

Subtract. Use addition to check your answer.

2.
```
   1,3 9 2
 −   2 3 8
```
Check:

3.
```
   3,2 9 8
 −   8 5 8
```
Check:

4.
```
   3,4 7 5
 − 1,2 6 7
```
Check:

Algebra Subtract to find the unknown.

5. $4,875 − $3,168 = ?

6. $6,182 − $581 = ?

7. 6,340 − 3,451 = ?

The unknown is _____.   The unknown is _____.   The unknown is _____.

Algebra Compare. Use >, <, or =.

8. 1,543 − 984 ◯ 5,193 − 4,893

9. 2,116 − 781 ◯ 5,334 − 3,999

Lesson 6 Subtract Four-Digit Numbers 167

Problem Solving

10. **Processes &Practices 2** **Use Algebra** Of the 2,159 pre-sold concert tickets, only 1,947 tickets were used. Write a number sentence to show how many tickets were not used.

    _____

11. Belinda is buying one of two cars. One costs $8,463 and the other costs $5,322. How much money would Belinda save if she bought the less expensive car?

    $ _____ − $ _____ = $ _____

Brain Builders

12. **Processes &Practices 2** **Reason** A group of students used 6,423 cans to create a sculpture. Another group made a sculpture using 2,112 cans. A third group made a sculpture with 4,279 cans. What is the difference in the greatest and least number of cans used for the sculptures? How do you know your answer is correct?

    _____
    _____
    _____

13. **Building on the Essential Question** Explain how subtracting four-digit numbers is like subtracting three-digit numbers. How is it different?

    _____
    _____
    _____
    _____

168 Chapter 3 Subtraction

Name

Lesson 6
Subtract Four-Digit Numbers

Homework Helper

Need help? connectED.mcgraw-hill.com

Find 4,453 − 2,474.

 Subtract ones.

Regroup 1 ten as 10 ones.

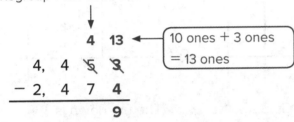

10 ones + 3 ones = 13 ones

 Subtract tens.

Regroup 1 hundred as 10 tens.

```
       14     ← 4 tens + 10 tens
   3   4  13     = 14 tens
4, 4   5   3
−2, 4   7   4
       7   9
```

Subtract hundreds and thousands.

Regroup 1 thousand as 10 hundreds.

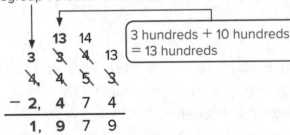

3 hundreds + 10 hundreds = 13 hundreds

Check

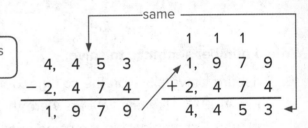

Addition shows the answer is correct.

So, 4,453 − 2,474 = 1,979.

Practice

1. Subtract. Use addition to check your answer.

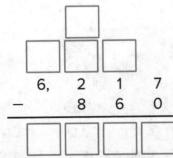

Check:

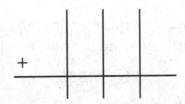

Lesson 6 My Homework 169

Subtract. Use addition to check your answer.

2.
```
    5, 9 6 3
  −    2 3 8
```
Check:
```
  +
```

3.
```
    5, 7 6 9
  −    9 4 1
```
Check:
```
  +
```

4.
```
    9, 8 7 1
  − 1, 2 1 4
```
Check:
```
  +
```

Algebra Subtract to find the unknown.

5. $3,958 − $1,079 = ?

6. $8,337 − $483 = ?

7. 6,451 − 2,378 = ?

The unknown is $_____.   The unknown is $_____.   The unknown is _____.

Brain Builders

Write a number sentence to solve.

8. Pittsburg University won the college football championship in 1937. They won again in 1976. How many years were there between championships? If they had won a championship the same number of years before 1937, what year would that have been?

9. **Processes & Practices 4** **Model Math** A library has 776 books about hobbies, 1,444 books about sports, and 1,814 books about animals. How many more books are there about sports and hobbies than there are about animals?

10. **Test Practice** How much less money did Selena's school raise this year at the pancake breakfast?

  Ⓐ $900   Ⓒ $1,905

  Ⓑ $905   Ⓓ $8,145

LAST YEAR
$4,525

THIS YEAR
$3,620

170 **Need more practice?** Download Extra Practice at connectED.mcgraw-hill.com

Name _____

Lesson 7
Subtract Across Zeros

ESSENTIAL QUESTION
How are the operations of subtraction and addition related?

Math in My World

Example 1

A large box of watermelons weighs 300 pounds. A smaller box weighs 134 pounds. What is the difference in the weight of the two boxes?

Find the unknown. 300 − 134 = ▪

Estimate 300 − 134 ⟶ 300 − 100 = ☐

1 Regroup.

```
  2 10
  3̸ 0̸ 0
−   1 3 4
```

Regroup 1 hundred as 10 tens.

2 Regroup once more. Then subtract by placing an X on the blocks.

```
        9
  2  1̸0  10
  3̸  0̸   0̸
−    1   3   4
  ☐  ☐   ☐
```

Regroup 1 ten as 10 ones.

Subtract the ones, tens, and hundreds.

Check Use _____ to check.

```
  3 0 0          1 6 6
−   1 3 4      + 1 3 4
  ☐              3 0 0
```

Addition shows the answer is correct.

_____ is close to the estimate _____.
Estimation shows the answer is reasonable.

So, 300 − 134 = _____ pounds. The unknown is _____.

Online Content at connectED.mcgraw-hill.com

Lesson 7 Subtract Across Zeros **171**

Use what you have learned about regrouping two times to regroup three times.

Example 2

A school bought music equipment for $5,004. The drums cost $2,815. How much money was spent on the other music equipment?

Find the unknown. $5,004 − $2,815 = ■

Estimate $5,004 − $2,815 ⟶ $5,000 − $2,800 = $2,200

1 Regroup 1 thousand, then 1 hundred, and 1 ten.

```
           9  9
     4   10 10  14
$   5,   0   0   4
−$  2,   8   1   5
```

2 Subtract starting at the ones place.

```
           9  9
     4   10 10  14
$   5,   0   0   4
−$  2,   8   1   5
─────────────────
$  ☐   ☐   ☐   ☐
```

Check

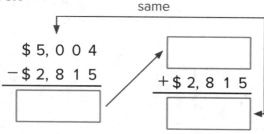

same

+ $ 2, 8 1 5

Addition shows the answer is correct.
Estimation shows the answer is reasonable.

So, $5,004 − $2,815 = _____. The unknown is _____.

Guided Practice

Subtract. Use addition to check.

1.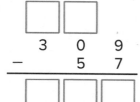
```
      3  0  9
   −     5  7
   ─────────
     ☐  ☐  ☐
```
Check:

+

2.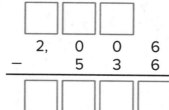
```
   2,  0  0  6
   −   5  3  6
   ──────────
     ☐ ☐ ☐ ☐
```
Check:

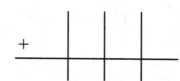

+

Talk MATH

Explain where you would start regrouping to find the difference in the problem 6,000 − 3,475.

Name ....

Independent Practice

Subtract. Use addition to check your answer.

3.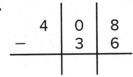

Check:

4.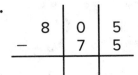

Check:

5.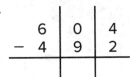

Check:

Algebra Subtract to find the unknown.

6. $9,006 − $7,474 = ■

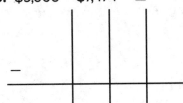

The unknown is $ _____.

7. 8,007 − 4,836 = ■

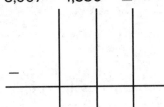

The unknown is _____.

8. $9,003 − $5,295 = ■

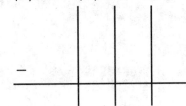

The unknown is _____.

9. 3,070 − 2,021 = ■

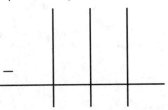

The unknown is _____.

10. 1,007 − 972 = ■

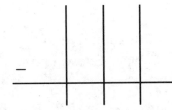

The unknown is _____.

11. 9,560 − 7,920 = ■

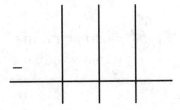

The unknown is _____.

Lesson 7 Add Three-Digit Numbers 173

Problem Solving

For Exercises 12 and 13, use the prices shown.

12. Anton purchased the bike. He gave the clerk one $100-bill, one $50-bill, and one $20-bill. How much change should Anton receive?

13. **Processes &Practices** **Draw a Conclusion** Suppose Anton decided to purchase the skates also, after he gave the clerk his money. How much more money does Anton need to give the clerk?

Brain Builders

14. **Processes &Practices** **Find the Error** Eva is solving the subtraction problem shown. Julian said 5,300 − 4,547 = 863. Find each mistake and correct it.

$$\begin{array}{r} 5,300 \\ -4,547 \\ \hline 1,853 \end{array}$$

15. **Building on the Essential Question** When might I need to regroup more than one time? Explain. Show an example to support your explanation.

174 Chapter 3 Subtraction

Name _____

Lesson 7
Subtract Across Zeros

Homework Helper

Need help? connectED.mcgraw-hill.com

Makenna won 3,000 tickets playing an arcade game. She used 1,872 tickets to buy a prize. How many tickets does she have left?

3,000 − 1,872 =   ← Find the unknown.

Estimate 3,000 − 1,900 = 1,100

1 Regroup 1 thousand, 1 hundred, and 1 ten.

```
      9 9
   2 10 10 10
    3, 0 0 0
  − 1, 8 7 2
```

2 Subtract, starting at the place farthest to the right.

```
      9 9
   2 10 10 10
    3, 0 0 0
  − 1, 8 7 2
    1, 1 2 8
```

Check

```
    1, 1 2 8
  + 1, 8 7 2
    3, 0 0 0
```

Addition shows the answer is correct.

1,128 is close to the estimate of 1,100. Estimation shows the answer is reasonable.

Practice

Subtract. Use addition to check.

1.
```
    5 0 7
  −   9 4
```
Check:

2.
```
    8 0 4
  − 6 6 7
```
Check:

3.
```
  4, 0 0 0
  −   9 6 9
```
Check:

Lesson 7 My Homework 175

Subtract. Use addition to check your answer.

4.
```
    7,0 0 5
  -   9 4 1
```
Check:

5.
```
      4 0 0
    - 1 5 3
```
Check:

6.
```
    3,0 0 0
  - 1,2 0 2
```
Check:

Algebra Subtract to find the unknown.

7. $3,008 − $1,053 = ■

■ = _____

8. 8,200 − 875 = ■

■ = _____

9. 9,001 − 3,860 = ■

■ = _____

Brain Builders

Processes & Practices 2 Understand Symbols Write a number sentence.

10. A bag holds 5,300 seeds. Brandon plants 790 seeds. Sharon plants 1,025 seeds. How many seeds are left?

11. Amy's book has 500 pages. She read 142 pages on Friday and 103 pages on Saturday. How many pages does Amy have left to read?

12. **Test Practice** Pete scored 1,400 points. Pete and Jocelyn compare their scores to Harrison's score. How much greater is Pete's difference than Jocelyn's difference?

   Ⓐ 64 points   Ⓒ 424 points

   Ⓑ 360 points   Ⓓ 488 points

Jocelyn 1,040   Harrison 976

176 **Need more practice?** Download Extra Practice at connectED.mcgraw-hill.com

Name _____

Fluency Practice

Processes & Practices 6

Subtract.

1. 94
  −13

2. 63
  −21

3. 84
  −73

4. 63
  −46

5. 42
  −27

6. 74
  −38

7. 58
  −29

8. 89
  −75

9. 71
  −56

10. 88
   −57

11. 73
   −59

12. 92
   −47

13. 422
   − 83

14. 111
   − 42

15. 299
   − 82

16. 604
   − 92

17. 476
   −229

18. 800
   −293

19. 493
   −310

20. 395
   −395

Online Content at connectED.mcgraw-hill.com

Fluency Practice 177

Name _____

Fluency Practice

Subtract.

1. 166
   − 34

2. 73
   −19

3. 91
   −51

4. 184
   − 78

5. 425
   −246

6. 661
   −487

7. 973
   −390

8. 451
   − 85

9. 730
   −692

10. 493
   −298

11. 671
   −479

12. 682
   −595

13. 625
   −263

14. 700
   −379

15. 990
   −372

16. 338
   − 70

17. 260
   − 99

18. 428
   −326

19. 511
   −300

20. 906
   −274

178 Chapter 3 Subtraction

Review

Chapter 3
Subtraction

Vocabulary Check

Choose the correct word(s) to complete each sentence.

difference   estimate   inverse operations

regroup   subtract   subtraction sentence

1. The answer to a subtraction problem is called the _____.

2. When I _____, I take away the lesser number from the greater number.

3. I can _____ to trade equal amounts when renaming a number.

4. When I do not need an exact answer I can _____ to find one that is close.

5. A number sentence in which one quantity is taken away from another quantity is a _____.

6. Two operations that can undo each other, as in addition and subtraction, are called _____.

Concept Check

Subtract mentally by breaking apart the smaller number.

7. 884 − 51 = _____

8. 283 − 171 = _____

9. 724 − 616 = _____

My Chapter Review 179

Estimate. Round each number to the indicated place value.

10. 765 − 121; tens

_____ − _____ = _____

11. 2,219 − 1,109; hundreds

_____ − _____ = _____

Subtract. Use addition to check your answer.

12.
```
    6 | 5 | 3
−   2 | 2 | 7
```
Check:
```
+
```

13.
```
    5 | 0 | 0
−   1 | 3 | 0
```
Check:
```
+
```

14.
```
    3, | 4 | 8 | 5
−   1, | 2 | 9 | 7
```
Check:
```
+
```

Algebra Subtract to find the unknown.

15. 608 − 45 = ■

The unknown is _____.

16. $3,568 − $639 = ■

The unknown is $_____.

17. 3,008 − 1,836 = ■

The unknown is _____.

Algebra Use addition to find each unknown.

18.
```
    5 | 8 | 0
−   ■ | ▲ | 0
    4 | 3 | 0
```
■ = _____
▲ = _____

19.
```
    1 | ■ | 5
−     | 4 | ▲
        8 | 6
```
■ = _____
▲ = _____

20.
```
    6, | 9 | 2 | 0
−   ■, |   | 1 | ▲ | 8
    4, | 7 | 2 | 2
```
■ = _____
▲ = _____

180 **Chapter 3** Subtraction

Problem Solving

21. An artist created a piece of art with 675 round glass beads. There were also 179 beads in the shape of a heart. Write a number sentence to find about how many more round beads there were.

_____

22. There are 365 days in one year. There were 173 sunny days this year. Write a number sentence to find the number of days that were not sunny.

_____

Brain Builders

23. In which month were fewer raffle tickets sold? How many fewer? Explain.

TV Raffle Ticket Sales		
Day	March	April
Saturday	$3,129	$4,103
Sunday	$3,977	$3,001

_____

_____

_____

24. Test Practice Students want to make 425 get well cards. So far, the boys have made 79 cards, and the girls have made 86 cards. How many more cards do they need to make?

　Ⓐ 240 cards　　　Ⓒ 270 cards

　Ⓑ 260 cards　　　Ⓓ 590 cards

Reflect

Chapter 3
Answering the
ESSENTIAL QUESTION

Use what you learned about subtraction to complete the graphic organizer.

ESSENTIAL QUESTION

How are the operations of subtraction and addition related?

Real World Problem

Models

Vocabulary

Reflect on the ESSENTIAL QUESTION Write your answer below.

_____

_____

_____

182 Chapter 3 Subtraction

Performance Task

Subtraction Race

A relay race will be held at the go-cart track among four teams. Each team will have three carts. Instead of keeping track of each team's finishing time, they will keep track of how many seconds each team is behind the winning team.

Show all your work to receive full credit.

Part A

Team 3 wins the race. Team 1 was 8 seconds behind the winner, and Team 4 was 26 seconds behind Team 1. Team 2 was 17 seconds behind Team 4. Find the finishing times of Team 3, Team 1, and Team 4 if Team 2 finished in 380 seconds. How many seconds ahead of Team 1 was Team 2?

Part B

Each team has three carts for the relay race, and the total time for each team is the sum of all three carts' times. For Team 1, the first cart finished in 108 seconds, and the second cart finished in 116 seconds. For Team 4, the first cart finished in 118 seconds and the second cart finished in 125 seconds. How many seconds did it take the third cart for Team 1 and the third cart for Team 4 to finish?

Part C

The first cart for Team 3 finished in 104 seconds and the second cart finished in 115 seconds. In how many fewer seconds than the third cart did the first cart finish?

Part D

The total time for all the first carts was 473 seconds and the total time for all the second carts was 474 seconds. Find the total time for all the teams by adding each team's finishing time. What was the total time for all the third carts?

Part E

Was the total time of the first carts less than or greater than the total time for the third carts? By how many seconds?

Chapter 4
Understand Multiplication

ESSENTIAL QUESTION

What does multiplication mean?

My Favorite Foods

Watch a video!

Name _____

MY Chapter Project

The Fruit Store

1. Each group member will create store "inventory" by choosing at least 1 fruit from the table below and drawing the chosen fruit on an index card. Each group member creates more than one index card of the fruit chosen.

2. As a class, create a price sign for the fruit. Prices are as follows:

Fruit	Price
Grapes	1¢
Plums	2¢
Apples	4¢
Oranges	5¢
Pineapples	10¢

3. When it is your turn to "shop," your group will become the "customer" and select a variety of fruits to "purchase." You will compute your bill using multiplication for the purchase price of more than one unit of a fruit. Then add the totals for each type of fruit purchased. Use the space below to show your work.

184 Chapter 4 Understand Multiplication

Name _____

Am I Ready?

Find each sum.

1. 2 + 2 + 2 + 2 = _____
2. 4 + 4 = _____
3. 5 + 5 + 5 = _____

4. 10 + 10 + 10 + 10 = _____
5. 0 + 0 + 0 = _____
6. 1 + 1 + 1 = _____

Write an addition sentence for each picture.

7.

_____ + _____ + _____ = _____

8.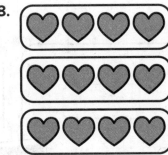

_____ + _____ + _____ = _____

Solve. Use repeated addition.

9. Larisa has 2 cups with 4 crackers in each cup. How many crackers does she have in all?

_____ crackers

10. On Monday and Tuesday, Lance rode his bike around the block 3 times each day. How many times in all did he ride his bike around the block?

_____ times

Shade the boxes to show the problems you answered correctly.

How Did I Do?

| 1 | 2 | 3 | 4 | 5 | 6 | 7 | 8 | 9 | 10 |

Online Content at connectED.mcgraw-hill.com

MY Math Words

Review Vocabulary

number sentence   repeated addition   sum

Making Connections
Use the review words to label each example of addition in the graphic organizer. Use the second column to draw an example of the addition.

Label It   **Draw It**

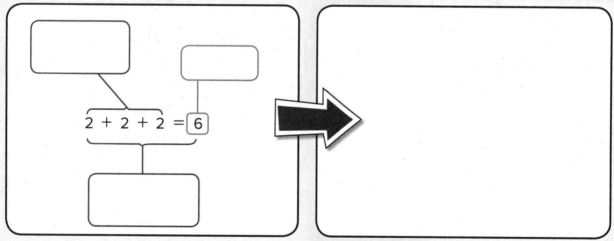

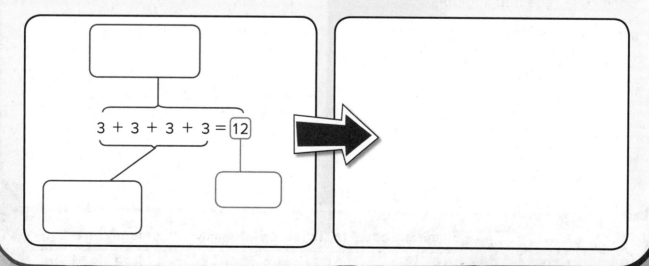

186 Chapter 4 Understand Multiplication

MY Vocabulary Cards

Processes & Practices

Lesson 4-3

array

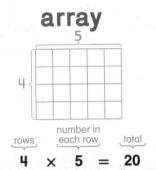

$\underbrace{4}_{\text{rows}} \times \underbrace{5}_{\text{number in each row}} = \underbrace{20}_{\text{total}}$

Lesson 4-6

combination

	pants	pants
shirt	shirt, pants	shirt, pants
shirt	shirt, pants	shirt, pants

$2 \times 2 = 4$

Lesson 4-3

Commutative Property of Multiplication

$3 \times 6 = 6 \times 3$

Lesson 4-1

equal groups

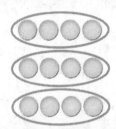

Lesson 4-2

factor

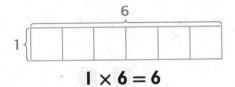

$1 \times 6 = 6$

Lesson 4-1

multiplication sentence

$3 \times 5 = 15$

Lesson 4-1

multiply (multiplication)

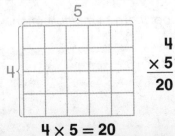

$\begin{array}{r} 4 \\ \times 5 \\ \hline 20 \end{array}$

$4 \times 5 = 20$

Lesson 4-2

product

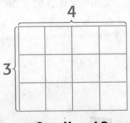

$3 \times 4 = 12$

Ideas for Use

- Group two or three related words. Add an unrelated word. Have another student identify which word is unrelated.
- Find pictures to show some of the words. Have a friend guess which word each picture shows.

A new set that has one item from each group of items.

What is the root word in *combination?* Write two other words using this root and different suffixes.

Objects or symbols displayed in rows of the same length and columns of the same length.

Disarray means "not orderly." Write the prefix in *disarray* and its meaning.

Groups with the same number of objects.

Draw an example of 4 equal groups.

The property that states that the order in which two numbers are multiplied does not change the product.

What part of the word commutative means *to go back and forth?*

A number sentence using the × sign.

Write an example of a multiplication sentence. Then write the sentence using words.

A number that divides a whole number evenly. Also a number that is multiplied by another number.

Write a new word you can make from *factor*. Include the definition.

The answer to a multiplication problem.

Write a number sentence with a product of 6.

An operation on two numbers to find their product.

Write the root word of *multiplication*.

MY Vocabulary Cards

Lesson 4-6

tree diagram

Food	Color	Combination
apple	red	apple, red
	green	apple, green
pepper	red	pepper, red
	green	pepper, green

Ideas for Use

- Group two or three related words. Add an unrelated word. Have another student identify which word is unrelated.
- Find pictures to show some of the words. Have a friend guess which word each picture shows.
- Use the blank cards to write your own vocabulary cards.

A diagram of all the possible outcomes of an event or series of events or experiments.

Explain how a *tree diagram* is like a tree.

_____

_____

MY Foldable

FOLDABLES Follow the steps on the back to make your Foldable.

How many combinations?

___ colors × ___ shapes = ___ combinations

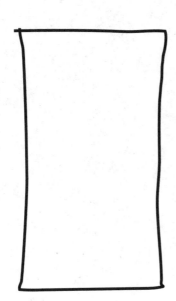 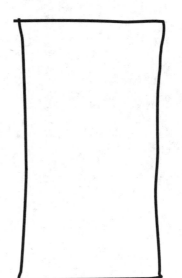 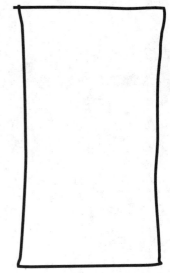

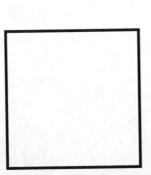 

 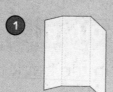 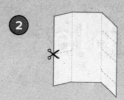 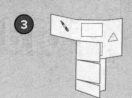

Name

Lesson 1 Hands On
Model Multiplication

ESSENTIAL QUESTION
What does multiplication mean?

Build It

When you have **equal groups,** you have the same number of objects in each group. Use repeated addition to find the total number of objects.

Find the total in 4 equal groups of 5.

1. Use connecting cubes to show 4 equal groups of 5 cubes. Draw the groups.

   There are _____ groups with _____ in each group.

2. Write the number of cubes in each group. Use repeated addition to complete the number sentence.

   _____ + _____ + _____ + _____ = _____

   The total in 4 groups of 5 is 20.

3. Record the number of groups, the number in each group, and the total from above.

   Explore other ways to group the 20 cubes equally.

Number of Groups	Number in Each Group	Total
4	5	20
10	2	20

Online Content at connectED.mcgraw-hill.com

Lesson 1   193

You can also use **multiplication** to find the total number of objects in equal groups. A number sentence with the symbol (×) is called a **multiplication sentence**. It means to **multiply**.

Try It

Shawn helped his mom bake cookies. He served 4 cookies on each plate. There are 2 plates. How many cookies did he serve?

Find the total of 2 plates of 4 cookies.

 Use counters to model the equal groups. Draw the groups.

 Use repeated addition to complete the number sentence.

_____ + _____ = _____

 Write a multiplication sentence to show 2 plates of 4 cookies, or 2 groups of 4.

_____ × _____ = _____
number of   number in   total
groups   each group

So, Shawn served _____ cookies.

Talk About It

1. **Processes & Practices 3** **Draw a Conclusion** How can addition help you find a total number of objects in equal groups?

_____

2. How did you find the total number of cubes in Step 2 of the first activity?

_____

3. Shawn counted a batch of cookies by finding 4 + 4 + 4. How could multiplication have helped him to find the total?

_____

What was the total? _____

Name ....

Practice It

Draw a model to find the total number.

4. 6 groups of 2 equals _____   **5.** 3 groups of 5 equals _____

6. 2 × 4 = _____   **7.** 1 × 7 = _____

Describe each set of equal groups.

8.

9.

_____ + _____ = _____   _____ + _____ + _____ = _____

_____ groups of _____ = _____   _____ groups of _____ = _____

10. 8 × 2 = _____   **11.** 5 × 5 = _____

_____ groups of _____ = _____   _____ groups of _____ = _____

Equally group the counters. Draw the equal groups.

12. set of 6 counters   **13.** set of 18 counters

Lesson 1 Hands On: Model Multiplication 195

Apply It

Processes & Practices 4 Model Math Draw to complete each model. Then complete each number sentence.

14. Tennis balls come in cans of 3. How many tennis balls are in 4 cans?

_____ + _____ + _____ + _____ = _____ tennis balls

15. Sam has 2 celery sticks. Each stick is topped with peanut butter and 4 raisins. How many raisins does Sam have in all?

_____ + _____ = _____ raisins

16. **Processes & Practices 2** Use Number Sense Mary bought a box of 6 crayons. Then she bought 3 more boxes. How many crayons did she buy in all? How much money did she spend in all?

_____ + _____ + _____ + _____ = _____ crayons; _____

Write About It

17. What does it mean to multiply?

_____

_____

196 Chapter 4 Understand Multiplication

Name _____

MY Homework

Lesson 1

Hands On: Model Multiplication

Homework Helper

Need help? connectED.mcgraw-hill.com

Gina put 3 scoops of frozen yogurt in each bowl. There are 6 bowls. How many scoops of frozen yogurt are there?

1. The model shows the total number of scoops.

   There are 6 bowls, and each has 3 scoops. There are 6 groups of 3.

2. Use repeated addition to find the total.

   3 + 3 + 3 + 3 + 3 + 3 = 18

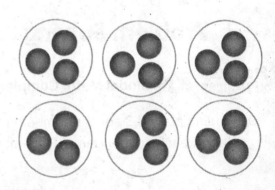

There are 18 scoops of frozen yogurt.

Practice

Draw a model to find the total number.

1. 2 groups of 8 equals _____

2. 5 groups of 7 equals _____

3. 6 × 4 = _____

4. 4 × 8 = _____

Describe each set of equal groups.

5.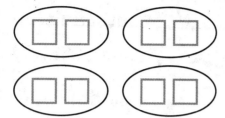

___ + ___ + ___ + ___ = ___

___ groups of ___ = ___

6.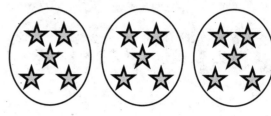

___ + ___ + ___ = ___

___ groups of ___ = ___

7. 6 × 5 = ___

___ groups of ___ = ___

8. 3 × 4 = ___

___ groups of ___ = ___

 Problem Solving

 Model Math Complete each number sentence.

9. Paulina played 3 soccer games on Saturday. She drank 1 juice box during each soccer game. How many juice boxes did she drink?

___ + ___ + ___ = ___ juice boxes

10. Daniel, Jamie, Molly, and Corey each have 4 books from the library. How many books do they have in all?

___ + ___ + ___ + ___ = ___ library books

Vocabulary Check

11. Choose the correct word(s) to complete the sentence below.

    equal groups   multiplication

    You can use _____ to find the total number of

    objects in _____ .

198 Chapter 4 Understand Multiplication

Name _____

Lesson 2
Multiplication as Repeated Addition

ESSENTIAL QUESTION
What does multiplication mean?

There are many ways to find the total when there are groups of equal objects.

Math in My World

Example 1

Gilberto made 4 small pizzas for his party. Each pizza had 5 slices of tomato. How many slices of tomato did Gilberto use to make 4 small pizzas?

Find how many slices of tomato there are in 4 groups of 5.

One Way Draw a picture.

 There are _____ groups. Draw 4 pizzas.

 There are _____ in each group.
Draw 5 slices of tomato on each pizza.

 Count. There are _____ slices of tomato.

Another Way Use repeated addition.
Write an addition sentence to show the equal groups.

_____ + _____ + _____ + _____ = _____

So, _____ groups of _____ is _____.

Gilberto used _____ slices of tomato.

Online Content at connectED.mcgraw-hill.com

Lesson 2 199

When you find the total of equal groups of objects, you **multiply**. The symbol (×) means to multiply. The numbers multiplied are **factors**. The result is the **product**.

Example 2

A honeycomb cell has 6 sides. How many sides do 5 separated honeycomb cells have?

Find 5 groups of 6.

One Way Write an addition sentence.

Use five 6-sided pattern blocks. Count the number of sides in all.

____ + ____ + ____ + ____ + ____ = ____

Another Way Write a multiplication sentence.

Find the unknown, or missing value.

So, ____ groups of 6 is ____. The unknown is ____ sides.

Guided Practice

1. Write an addition sentence and a multiplication sentence.

____ + ____ + ____ + ____ = ____

____ × ____ = ____

Talk MATH

Can you write 2 + 3 + 4 = 9 as a multiplication sentence? Explain.

200 Chapter 4 Understand Multiplication

Name _____

Independent Practice

Write an addition sentence and a multiplication sentence for each.

2.

___ + ___ + ___ = ___

___ × ___ = ___

3.

___ + ___ = ___

___ × ___ = ___

4.

___ + ___ + ___ + ___ + ___ + ___ = ___

___ × ___ = ___

5.

___ + ___ + ___ + ___ + ___ + ___ + ___ = ___

___ × ___ = ___

Draw a picture to find the total. Write a multiplication sentence.

6. 6 groups of 5

7. 8 groups of 4

___ × ___ = ___

___ × ___ = ___

Algebra Multiply to find the unknown product.

8. 3 × 5 = ■

9. 5 × 2 = ■

10. 3 × 3 = ■

The unknown is ___.

The unknown is ___.

The unknown is ___.

Lesson 2 Multiplication as Repeated Addition

Problem Solving

Algebra Write a multiplication sentence with a symbol for the unknown. Then solve.

11. Adriano bought 3 boxes of paints. Each box has 8 colors. What is the total number of colors?

    $3 \times 8 = \blacksquare$

    $3 \times 8 =$ _____ colors

12. Three boys each have 5 balloons. How many balloons do they have all together?

    _____ × _____ = $\blacksquare$

    _____ × _____ = _____ balloons

Brain Builders

13. **Processes & Practices 4** **Model Math** Write a real-world problem for the model. Write a multiplication sentence to find the total.

_____

_____

14. **Processes & Practices 2** **Use Number Sense** What is 9 more than 5 groups of 3? How many groups of 4 can you make with that number?

_____

15. **Building on the Essential Question** How are multiplication and repeated addition alike? Give an example.

_____

_____

Name _____

Lesson 2

Multiplication as Repeated Addition

Homework Helper

Need help? connectED.mcgraw-hill.com

Dani will put 2 forks at each of the 8 table settings. How many forks does she need in all?

Find 8 groups of 2.

Write an addition sentence to show the equal groups.

2 + 2 + 2 + 2 + 2 + 2 + 2 + 2 = 16

Write a multiplication sentence to show 8 groups of 2.

8 × 2 =   ← Find the unknown.

8 × 2 = 16

So, 8 groups of 2 is 16. The unknown is 16 forks.

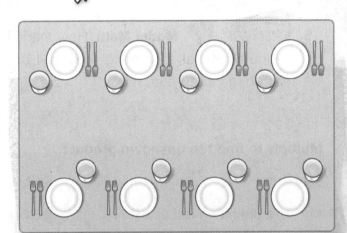

Practice

Write an addition sentence and a multiplication sentence for each.

1.

6 + 6 + ___ + ___ + ___ = ___

___ × ___ = ___

2.

___ + ___ + ___ + ___ = ___

___ × ___ = ___

Lesson 2 My Homework 203

Problem Solving

Draw a picture to find the total. Write a multiplication sentence.

3. 7 groups of 1 green grape

4. 9 groups of 2 square crackers

_____ × _____ = _____   _____ × _____ = _____

5. **Processes & Practices** **Model Math** How many buttons does Leonora have all together if she has 4 bags of buttons and each bag has 10 buttons?

_____ × _____ = _____

Multiply to find the unknown product.

6. $8 \times 3 = \blacksquare$

   The unknown is _____.

7. $4 \times 3 = \blacksquare$

   The unknown is _____.

Vocabulary Check

Use the correct word(s) and the number sentence $6 \times 8 = 48$ to solve.

equal groups   repeated addition   multiply   factors   product

8. The number 48 is the _____.

9. The symbol × tells you to _____.

10. The numbers 6 and 8 are the _____.

11. $8 + 8 + 8 + 8 + 8 + 8 = 48$ shows _____.

12. 6×8 means 6 _____ of 8.

Brain Builders

13. **Test Practice** Sam and Bo are washing windows. Each boy washes 5 windows in each of 7 rooms. How many windows do they wash?

    Ⓐ 2 windows   Ⓒ 35 windows

    Ⓑ 12 windows   Ⓓ 70 windows

204 **Need more practice?** Download Extra Practice at connectED.mcgraw-hill.com

Name _____

Lesson 3
Hands On
Multiply with Arrays

ESSENTIAL QUESTION
What does multiplication mean?

Draw It

An **array** has rows of equal length and columns of equal length.

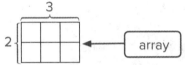

1 Make an array on a piece of paper. Arrange the tiles in 4 rows with 3 tiles in each row. Draw the array.

2 Count. What is the total number of tiles? _____

3 Turn your paper. There are now _____ rows with _____ tiles in each row.

Draw what the array looks like now.

4 Count. What is the total number of tiles? _____

So, there are the same number of tiles, _____, if you turn the array.

Online Content at connectED.mcgraw-hill.com

The **Commutative Property of Multiplication** states that the order in which numbers are multiplied does not change the product.

Try It

1. Use tiles to make an array on paper that has 5 rows of 2 tiles. Draw the array.

Write an addition sentence to show equal rows.

____ + ____ + ____ + ____ + ____ = ____

Write a multiplication sentence to represent the array.

rows   number in each row   total

____ × ____ = ____

2. Turn the array the other way. There are now ____ rows of ____ tiles. Draw the array.

Write an addition sentence to show equal rows.

____ + ____ = ____

Write a multiplication sentence to represent the array.

rows   number in each row   total

____ × ____ = ____

Talk About It

1. Processes & Practices 2 **Stop and Reflect** What is the connection between repeated addition and an array?

_____

2. How can you use an array to model the Commutative Property?

_____

3. List 3 everyday objects that are arranged in an array.

_____

206   Chapter 4 Understand Multiplication

Name

Practice It

Draw an array to find the product.

4. $4 \times 2 =$ _____

5. $3 \times 5 =$ _____

Write an addition sentence and a multiplication sentence to show equal rows.

6. 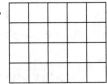   ___ + ___ + ___ + ___ = ___

___ × ___ = ___

7.   ___ + ___ + ___ = ___

___ × ___ = ___

8.   ___ + ___ = ___

___ × ___ = ___

9. Use the Commutative Property of Multiplication to write another multiplication sentence for Exercise 8.

___ × ___ = ___

10. Describe a real-world situation for Exercise 6.

_____

_____

Lesson 3 Hands On: Multiply with Arrays **207**

Apply It

11. Marcos has 3 sheets of stickers. Each sheet has 4 stickers on it. Write a multiplication sentence to find how many stickers he has in all.

_____ × _____ = _____ stickers

12. Circle the picture that does not represent an array. Explain.

_____
_____

13. Processes &Practices 1 **Keep Trying** Draw an array to find the unknown number in the multiplication sentence 6 × ■ = 18. What is the unknown?

Write About It

14. How can I use an array to model multiplication?

_____
_____

Name _____

Lesson 3

Hands On:
Multiply with Arrays

Homework Helper   Need help? connectED.mcgraw-hill.com

James discovered that a sheet of stamps is in an array. The stamps are arranged in 6 equal rows of 3.

Write an addition sentence to show equal rows.

$3 + 3 + 3 + 3 + 3 + 3 = 18$

Write a multiplication sentence to represent the array.

$$\underbrace{6}_{\text{rows}} \times \underbrace{3}_{\text{number in each row}} = \underbrace{18}_{\text{total}}$$

James turns the sheet of stamps the other way. There still are 18 stamps. Only now, there are 3 equal rows of 6.

$$\underbrace{3}_{\text{rows}} \times \underbrace{6}_{\text{number in each row}} = \underbrace{18}_{\text{total}}$$

This is the Commutative Property of Multiplication.

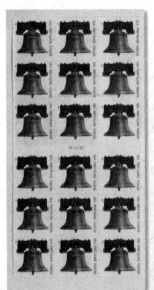

Practice

Draw an array to find the product.

1. $5 \times 7 =$ _____

2. $6 \times 5 =$ _____

Lesson 3 My Homework 209

Write an addition sentence and a multiplication sentence to show equal rows.

3.

_____

_____

4.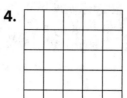

_____

_____

Vocabulary Check

5. Draw two arrays to model 2 × 3 = 6. Use the arrays to show the meaning of the Commutative Property of Multiplication.

Problem Solving

6. **Processes &Practices 4** **Model Math.** Suki's watercolor set has 3 rows of paint. There are 8 colors in each row. Write a multiplication sentence to find the total number of colors in the set.

_____

7. A checkerboard has 8 rows, with 8 squares in each row. Write a multiplication sentence to find the total number of squares.

_____

Name _____

Lesson 4
Arrays and Multiplication

ESSENTIAL QUESTION
What does multiplication mean?

An **array** is a group of objects arranged in equal numbered rows and equal numbered columns. Arrays can help you multiply.

Example 1

Mrs. Roberts baked a batch of bagels. She arranged the bagels in 3 equal rows of 4 bagels each on the cooling rack. How many bagels did she bake?

Find the total number of bagels. Use counters to model the array. Draw the array.

_____ rows of _____ = ▪

(Find the unknown.)

You can use repeated addition or multiplication to find the unknown.

One Way Add.   _____ + _____ + _____ = _____   ← addition sentence

Another Way Multiply.   _____ × _____ = _____   ← multiplication sentence

So, _____ rows of _____ is _____ or _____ × _____ = _____.

The unknown is _____. Mrs. Roberts baked 12 bagels.

Online Content at 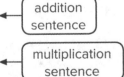 connectED.mcgraw-hill.com

Lesson 4 211

Example 2

One page of Elsa's photo album is shown. Write two multiplication sentences to find how many photos are on the page.

 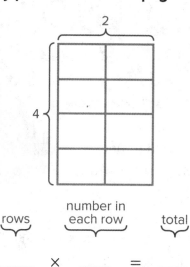 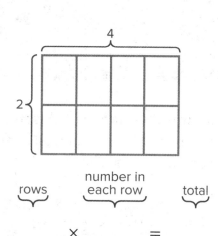

rows × number in each row = total   rows × number in each row = total

___ × ___ = ___   ___ × ___ = ___

Key Concept   Commutative Property

Words	The **Commutative Property of Multiplication** says the order in which numbers are multiplied does not change the product.
Examples	4 × 2 = 8 2 × 4 = 8
	factor factor product factor factor product

Guided Practice

Write an addition sentence and a multiplication sentence to show equal rows.

1.   ___ + ___ = ___
     ___ × ___ = ___

What other operation uses the Commutative Property? Explain.

2. ___ + ___ = ___

   ___ × ___ = ___

212 Chapter 4 Understand Multiplication

Name

Independent Practice

Write an addition sentence and a multiplication sentence to show equal rows.

3.

___ + ___ + ___ = ___

___ × ___ = ___

4.

___ + ___ = ___

___ × ___ = ___

5.

___ + ___ + ___ + ___ = ___

___ × ___ = ___

6.

___ + ___ + ___ = ___

___ × ___ = ___

7.

___ + ___ = ___

___ × ___ = ___

8.

___ + ___ = ___

___ × ___ = ___

Use the Commutative Property of Multiplication to find each missing number.

9. 5 × 2 = ___

   2 × ___ = 10

10. ___ × 5 = 15

    ___ × 3 = 15

11. 3 × ___ = 27

    9 × 3 = ___

12. Hope drew the array at the right. Write a multiplication sentence to represent the model.

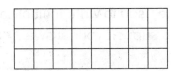

___ × ___ = ___

Lesson 4 Arrays and Multiplication 213

Problem Solving

For Exercises 13 and 14, draw an array to solve. Then write two multiplication sentences.

13. Bailey made a 3 by 4 array with her crackers. How many crackers does she have?

14. There are 4 waiters serving 5 tables each. How many tables do the waiters have all together?

Brain Builders

15. **Reason** Why do you sometimes have only one multiplication sentence for an array? What shape do these arrays take? Give an example.

16. **Find the Error** Alyssa is using the numbers 2, 3, and 6 to show the Commutative Property. Find and correct her mistake.

    $3 \times 2 = 6$ so, $6 \times 3 = 2$

17. **Building on the Essential Question** How can the Commutative Property be used to write multiplication sentences? Provide an example to support your answer.

214 Chapter 4 Understand Multiplication

Name

MY Homework

Lesson 4
Arrays and Multiplication

Homework Helper

Need help? connectED.mcgraw-hill.com

The pumpkins in a patch are arranged in rows with an equal number in each row. How many pumpkins are in the patch?

Write an addition sentence and a multiplication sentence to show equal rows.

6 + 6 + 6 = 18   3 × 6 = 18

The Commutative Property of Multiplication allows you to change the order of the factors to write another multiplication sentence, 6 × 3 = 18.

There are 18 pumpkins in the pumpkin patch.

Practice

Write an addition sentence and a multiplication sentence to show equal rows.

1.

2.

___ + ___ + ___ = ___

___ × ___ = ___

___ + ___ + ___ = ___

___ × ___ = ___

Lesson 4 My Homework 215

Use the Commutative Property of Multiplication to find each missing number.

3. $3 \times 2 = 6$ ____ $\times 3 = 6$

4. $6 \times 4 = 24$ $4 \times$ ____ $= 24$

5. $8 \times 6 = 48$ $6 \times 8 =$ ____

6. $5 \times 2 = 10$ ____ $\times 5 = 10$

Brain Builders

Draw an array to solve. Then write two multiplication sentences.

7. Bottles of syrup are arranged in 4 rows of 5 bottles each. Chris adds 2 more bottles to each row. How many bottles of syrup are there in all?

8. A parking lot had 4 rows of 10 spaces. Two more rows were added after the parking lot was expanded. How many parking spaces are there now?

Vocabulary Check

9. How can you use an array to show the Commutative Property?

10. **Test Practice** Jill writes $6 \times \boxed{} = 42$. Which number sentence represents the Commutative Property of Multiplication?

Ⓐ $3 \times 6 = 18$

Ⓒ $4 \times 6 = 24$

Ⓑ $7 \times 6 = 42$

Ⓓ $6 + 7 = 13$

Check My Progress

Vocabulary Check

Write the letter of the word that matches each definition or example.

A array

B Commutative Property of Multiplication

C equal groups

D factors

E multiplication

F multiplication sentence

G multiply

H product

J repeated addition

_____ 1. $2 + 2 + 2 + 2 = 8$

_____ 2. $5 \times 3 = 15$

_____ 3. $2 \times 4 = 8$

_____ 4. This symbol $\times$ means to _____.

_____ 5. An operation used to find the total number in equal groups.

_____ 6. $5 \times 3 = 15$

_____ 7. $3 \times 2 = 6$ $2 \times 3 = 6$

_____ 8.

_____ 9.

Concept Check

Algebra Circle equal groups. Find the unknown.

10. 2 groups of 4 = ■

    2 groups of 4 = _____

11. 4 groups of 5 = ■

    4 groups of 5 = _____

Check My Progress 217

Use repeated addition to multiply.

12. 6 × 2 = ____

____ + ____ + ____ +

____ + ____ + ____ = ____

13. 2 × 5 = ____

____ + ____ = ____

Write an addition sentence and a multiplication sentence to show equal rows.

14.

____ + ____ + ____ = ____

____ × ____ = ____

15.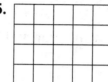

____ + ____ + ____ + ____ = ____

____ × ____ = ____

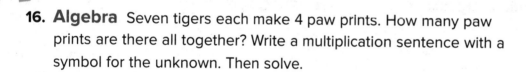

Brain Builders

16. **Algebra** Seven tigers each make 4 paw prints. How many paw prints are there all together? Write a multiplication sentence with a symbol for the unknown. Then solve.

17. Student tickets cost $4. Adult tickets cost $2 more. How much will 4 adult tickets cost? Draw an array to solve. Then write two multiplication sentences.

18. **Test Practice** Which pair of number sentences is related to the addition sentence 5 + 5 + 5 = 15?

    Ⓐ 3 × 5 = 15; 5 × 3 = 15
    Ⓑ 5 × 3 = 15; 3 + 5 = 8
    Ⓒ 15 − 5 = 10; 5 × 10 = 50
    Ⓓ 5 + 3 = 8; 8 − 5 = 3

218 **Chapter 4** Understand Multiplication

Name _____

Lesson 5
Problem-Solving Investigation
STRATEGY: Make a Table

ESSENTIAL QUESTION
What does multiplication mean?

Learn the Strategy

Selma bought 3 shorts and 2 shirts. Laura bought 4 shorts and 2 shirts. How many different shirt and shorts outfits can each girl make?

1 Understand
What facts do you know?

_____

What do you need to find?

How many different _____ and _____ outfits they each can make.

2 Plan
Organize the information in a table of columns and rows.

3 Solve
Make a table for each girl. List the possible shirt and shorts outfits.

Selma	Shirt 1	Shirt 2
Shorts A	A1	A2
Shorts B	B1	B2
Shorts C	C1	C2

Laura	Shirt 1	Shirt 2
Shorts A	A1	A2
Shorts B	B1	B2
Shorts C	C1	C2
Shorts D	D1	D2

So, Selma can make _____ outfits, and Laura can make _____.

4 Check
Does your answer make sense? Explain.

_____

Online Content at connectED.mcgraw-hill.com

Lesson 5 219

Practice the Strategy

How many lunches can Malia make if she chooses one main dish and one side dish?

Main Dishes
- tacos
- grilled cheese sandwich
- noodles

Side Dishes
- fruit
- soup
- veggies

1 Understand

What facts do you know?

_____

_____

What do you need to find?

_____

_____

2 Plan

_____

_____

3 Solve

4 Check

Does your answer make sense? Explain.

_____

Name ....

Apply the Strategy

Solve each problem by making a table.

1. Trey can choose one type of bread and one type of meat for his sandwich. How many different sandwiches can Trey make?

	Turkey	Chicken
Wheat		
White		

Trey can make _____ sandwiches.

Brain Builders

2. The students in Mr. Robb's class are designing a flag. The flag's background can be gold, red, or green. The flag can have either a blue or a purple stripe. Color all the possible flags.

	Blue	Purple
Gold		
Red		
Green		

They can design _____ flags.

If orange could be used as a background, they could design _____ more flags.

3. **Processes & Practices 4** **Model Math** Tracy has a picture of her mom, a picture of her dad, and a picture of her dog. She has a black frame and a white frame. What is the question? Solve.

Lesson 5 Problem-Solving Investigation 221

Review the Strategies

Use any strategy to solve each problem.
- Use the four-step plan.
- Determine reasonable answers.
- Make a table.

4. Amber has coins in a jar. The sum of the coins is 13¢. What are the possible groups of coins Amber could have?

5. Solana buys 2 bags of salad mix for $8 and 3 pounds of fresh vegetables for $9. She gives the cashier $20. How much change will she receive?

6. **Processes & Practices 6** **Be Precise** Mr. Grow has 12 tomato plants arranged in 2 rows of 6. List 2 other ways that Mr. Grow could arrange his 12 tomato plants in equal rows. Explain to a classmate how you got your answer.

7. **Processes & Practices 5** **Use Math Tools** One campsite has 3 tents with 5 people in each tent. Another campsite has 3 tents with 4 people in each tent. How many campers are there in all? Draw arrays to solve.

222 Chapter 4 Understand Multiplication

Name _____

MY Homework

Lesson 5

Problem Solving:
Make a Table

Homework Helper

Need help? connectED.mcgraw-hill.com

Jane's new bike can have hand brakes or foot brakes. The bike can be silver, blue, black, or purple. How many possible bikes are there?

1 Understand

There are 2 types of brakes: hand brakes or foot brakes.
There are 4 color choices: silver, blue, black, or purple.

I need to find the number of possible bikes.

2 Plan

Make a table.

	Silver	Blue	Black	Purple
Hand brakes	Hand/Silver	Hand/Blue	Hand/Black	Hand/Purple
Foot brakes	Foot/Silver	Foot/Blue	Foot/Black	Foot/Purple

3 Solve

There are 8 possible bikes.

4 Check

Multiply 2 types of brakes by 4 color choices. $4 \times 2 = 8$

Problem Solving

1. Solve the problem by making a table.

   Claudio will decorate his bedroom. He can choose tan, blue, or gray paint and striped or plaid curtains. How many ways can he decorate his room with different paint and curtains?

	tan, (t)	blue, (b)	gray, (g)
striped, (s)			
plaid, (p)			

Solve each problem by making a table.

2. Jimmy has a number cube labeled 1 through 6, and a penny. How many different ways can the cube and penny land with one roll of the cube and one flip of the penny?

	1	2	3	4	5	6
heads, (h)						
tails, (t)						

Brain Builders

3. Archie earns $6 each week for doing his chores. He puts all but $2 into a savings account. How much money will Archie save in 2 months if there are 4 weeks in each month?

	Week 1	Week 2	Week 3	Week 4
Month 1				
Month 2				

4. **Processes &Practices 7 Identify Structure** Abigail has a green, yellow, and purple shirt to match with either a white, black, or red pair of pants. How many different shirt and pants outfits can she make?

	pants, (w)	pants, (b)	pants, (r)
shirt, (g)			
shirt, (y)			
shirt, (p)			

How many outfits would be possible if Abigail had only 2 shirts and 2 pair of pants? Explain.

224 **Need more practice?** Download Extra Practice at connectED.mcgraw-hill.com

Name _____

Lesson 6
Use Multiplication to Find Combinations

ESSENTIAL QUESTION
What does multiplication mean?

When you make a **combination**, you make a new set that has one item from each group of items.

Math in My World

Example 1

Amos' team has 3 jersey colors: green, red, and yellow. They can wear orange or black shorts. Find all of the jersey and short combinations for the team.

1. Color the first jersey green, the second one red, and the last yellow.

2. Color 1 pair of shorts orange and 1 pair of shorts black below each jersey.

Combinations	1	2		4	5	
Jersey Colors	GREEN	GREEN	RED	RED		
Shorts Colors	ORANGE	BLACK	ORANGE		ORANGE	

Write a multiplication sentence.

_____ × _____ = ▨   ← Find the unknown.
jersey colors shorts colors combinations

_____ × _____ = _____

So, there are _____ jersey and shorts combinations possible.

Online Content at connectED.mcgraw-hill.com

Lesson 6   225

Another way to find combinations is to use a tree diagram.
A **tree diagram** uses "branches" to show all possible combinations.

Example 2

What are all the possible fruit sorbet combinations if you choose one flavor and one fruit to add in?

Complete the tree diagram.

Flavors	Fruits	Combinations
mango	banana — berries — peach —	mango, _____ mango, berries _____, peach
strawberry	banana — berries — peach —	_____, banana _____, berries _____, _____
vanilla	banana — berries — peach —	_____, _____ _____, _____ _____, _____

Check Multiply to find the number of possible combinations.

_____ flavors × _____ fruits = _____ fruit sorbet combinations

So, there are _____ possible fruit sorbet combinations.

Guided Practice

1. Refer to Example 2. How would the possible number of combinations change if one more flavor was added? Write the multiplication sentence.

   _____ × _____ = _____

Explain how a tree diagram helps you find all the possible combinations without repeating any.

226 Chapter 4 Understand Multiplication

Name ....

Independent Practice

Find all the possible combinations. Write a multiplication sentence to check.

2. Jackie is playing a card game with triangles and circles. Each shape can be blue, red, yellow, or green. How many different cards are there?

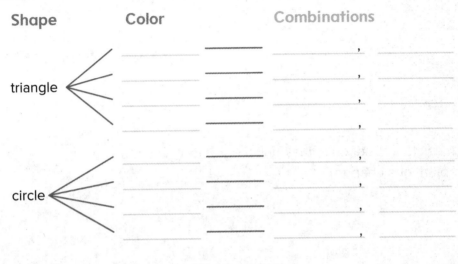

_____ × _____ = _____ different cards

3. List all of the 2-digit numbers that can be made with 3 or 4 as the tens digit and 1, 6, 7, 8, or 9 as the ones digit.

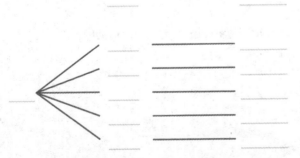

_____ × _____ = _____ numbers

Lesson 6 Use Multiplication to Find Combinations **227**

Problem Solving

Write a multiplication sentence to solve the problem.

4. Madison needs to choose 1 breakfast item and 1 drink. Find the number of possible combinations.

    _____ × _____ = _____ combinations

Breakfast Menu	Drink Menu
(pancakes)	(milk)
(eggs and bacon)	(juice)

    Suppose hot chocolate was added to the drink menu. How would the number of combinations change?

    _____

    _____

Brain Builders

5. **Processes & Practices 5** **Use Math Tools** Write a real-world combination problem for the multiplication sentence 4 × 2 = ■. Ask a classmate to solve with a tree diagram. Then find the unknown.

    _____

    _____

6. **Building on the Essential Question** How can multiplication help to find combinations? Give an example.

    _____

    _____

    _____

Name _____

Lesson 6
Use Multiplication to Find Combinations

Homework Helper

Need help? connectED.mcgraw-hill.com

Lucia's three dogs have red, purple, blue, green, and orange collars that they take turns wearing. Find the number of possible dog and collar combinations.

Show all of the possible combinations.

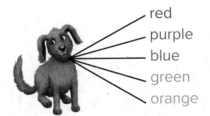
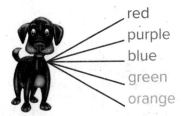
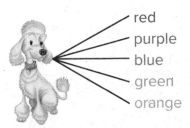

There are 3 dogs and 5 collar colors.
3 × 5 = 15 possible combinations

Practice

1. Diana can take 1 pencil and 1 eraser to school. Her choices are shown. How many different pencil and eraser combinations are there? Complete the tree diagram.

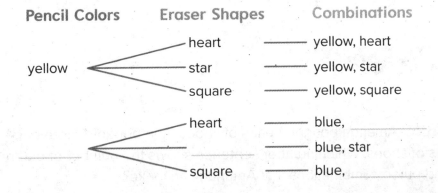

____ × ____ = ____ combinations

Lesson 6 My Homework 229

2. **Processes & Practices 7** **Identify Structure** For a snack, Randy can choose from carrots, peanuts, or popcorn. He can have water or juice to drink. How many snack and drink combinations are there? Complete the tree diagram. Write a multiplication sentence.

Snack   Drink   Combinations

____ × ____ = ____ combinations

Vocabulary Check

3. Write the correct vocabulary word(s) in each space to complete the sentence.

combination   tree diagram

Each branch of a _____ shows a possible

_____ of items.

Brain Builders

4. **Test Practice** Amanda bought 4 pairs of shoes. She bought 1 more purse than pairs of shoes. Which number sentence shows the number of different shoes and purse combinations that Amanda can make?

Ⓐ 4 + 5 = 9

Ⓒ 4 + 4 + 4 + 4 = 16

Ⓑ 4 × 1 = 4

Ⓓ 4 × 5 = 20

Review

Chapter 4
Understand Multiplication

Vocabulary Check

Use the word bank to complete each sentence.

array
equal groups
multiplication sentence
tree diagram
combination
factor
multiply
Commutative Property
multiplication
product

1. An arrangement of objects into rows of equal length and columns of equal length is a(n) _____.

2. The answer to a multiplication problem is the _____.

3. _____ is the operation of two numbers that can be thought of as repeated addition.

4. A number multiplied by another number is a _____.

5. You can put equal groups together to _____.

6. The _____ says the order in which numbers are multiplied does not change the product.

7. A _____ uses "branches" to show all possible combinations.

8. When you make a _____ of items, you make a new set that has one item from each group.

9. When you have _____, you have the same number of objects in each group.

Concept Check

Write an addition and a multiplication sentence to show equal rows.

10.

___ + ___ + ___ = ___

___ × ___ = ___

11.

___ + ___ = ___

___ × ___ = ___

Write two multiplication sentences for each array.

12.

___ × ___ = ___

___ × ___ = ___

13.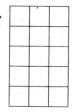

___ × ___ = ___

___ × ___ = ___

14. Find the possible combinations of one yogurt and one topping. Complete the tree diagram. Write a multiplication sentence to check.

Sundaes	
Yogurt	Toppings
strawberry	granola
peach	strawberries
vanilla	

Yogurt   Topping

strawberry
                 _____
                 strawberries

                 granola
                 _____

                 _____
                 _____

___ × ___ = ___ combinations

232 Chapter 4 Understand Multiplication

Problem Solving

15. There are 3 rows of 4 muffins. How many muffins all together? Write two multiplication sentences.

_____ × _____ = _____

_____ × _____ = _____

Brain Builders

16. Toya finishes reading a book every 3 days. How many days does it take her to read 7 books? How many more days will it take Toya to read 9 books? Complete the table to solve.

Days	Books
3	1
6	
12	
	5
18	6
21	

So, it takes her _____ days to read 7 books.

It will take her _____ more days to read 9 books.

17. Test Practice Timmy downloaded 3 pop songs and 2 country songs each day for five days. How many songs did he download during the 5 days all together?

Ⓐ 5 songs
Ⓑ 10 songs
Ⓒ 25 songs
Ⓓ 35 songs

Reflect

Chapter 4

Answering the
ESSENTIAL QUESTION

Use what you learned about multiplication to complete the graphic organizer.

- Draw Equal Groups
- Real-World Problem
- Vocabulary
- Draw an Array

ESSENTIAL QUESTION
What does multiplication mean?

Now reflect on the ESSENTIAL QUESTION. Write your answer below.

Name _____   Date _____

Score _____

Performance Task

Fashion Designer

Sophie loves clothes and colors. Sometimes she designs outfits of all one color. Sometimes she designs outfits of different colors that go well together.

Show all your work to receive full credit.

Part A

Sophie has designed eight different colors of shirts in four different styles. How many shirts has she designed in all? Draw a model to support your answer.

Part B

Sophie has designed two scarves of each color: black, blue, red, green, yellow, pink, and purple. How many scarves has she designed? Write an addition sentence and a multiplication sentence for the number of scarves she has designed.

Online Content at connectED.mcgraw-hill.com   Performance Task   234PT1

Part C

Sophie chooses one color of shirt and one color of scarf to model each day. How many different color combinations of shirts and scarves can she model? Write a multiplication sentence to support your answer.

Part D

Sophie is thinking of designing either a new shirt or a new scarf. How many more possible combinations of shirts and scarves will Sophie be able to model if she designs one new color of shirt? How many more possible combinations will she be able to model if she designs one new color of scarf?

Part E

This week, Sophie loaned another designer three colors of shirts. The other designer loaned Sophie two different colors of scarves. How many combinations of shirts and scarves can Sophie model this week? Is it more combinations or fewer combinations than in **Part C**?

Chapter

5 Understand Division

ESSENTIAL QUESTION

What does division mean?

Careers in OUR World

Watch a video!

Name
....

MY Chapter Project

Division Classroom Bulletin Board

1. As a class, you will brainstorm ideas for the division bulletin board.

2. Division concept your group will display: _____

3. As a group, discuss how you will display your concept. Sketch your group's plans in the space below.

4. Think of a question your classmates can answer by using your group's display on the division bulletin board. Sample question: *Explain how division and subtraction are related.*

Question: _____

_____

_____

Name _____

Write two multiplication sentences for each array.

1. 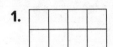   2.

_____ _____   _____ _____

Identify a pattern. Then find the missing numbers.

3. 30, 25, 20, _____, _____, 5   Pattern: _____

4. 12, _____, 8, _____, 4, 2   Pattern: _____

5. 55, 45, 35, _____, 15, _____   Pattern: _____

Draw the counters in the circles to make equal groups.

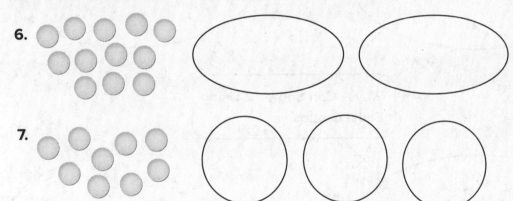

8. Colton made 15 party invitations. His brother made 9 invitations. Write a subtraction sentence to find how many more invitations Colton made.

_____

9. Mrs. Jones has 21 pencils. She gives 2 pencils to Carter and 2 pencils to Mandy. Write a subtraction sentence to find how many pencils she has left.

_____

Shade the boxes to show the problems you answered correctly.

How Did I Do? | 1 | 2 | 3 | 4 | 5 | 6 | 7 | 8 | 9 |

Online Content at connectED.mcgraw-hill.com

Name
_____

MY Math Words

Review Vocabulary

| array | equal groups | pattern | repeated addition |

Making Connections
Use the review vocabulary to describe each example in the graphic organizer.

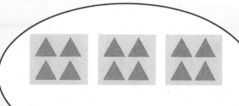

_____

[array image]

_____

Ways I Can Show 12 Using Equal Groups

12, 8, 4, 0

_____

4 + 4 + 4

_____

How are the examples similar? How are the examples different?

_____

_____

238 Chapter 5 Understand Division

MY Vocabulary Cards

Processes & Practices

Lesson 5-1

divide (division)

$12 \div 3 = 4$

Lesson 5-4

dividend

$15 \div 3 = 5$

Lesson 5-1

division sentence

$15 \div 3 = 5$

Lesson 5-4

divisor

$15 \div 3 = 5$

Lesson 5-5

fact family

$9 \times 3 = 27 \quad 27 \div 9 = 3$

$3 \times 9 = 27 \quad 27 \div 3 = 9$

Lesson 5-5

inverse operations

$2 \times 5 = 10$

$10 \div 2 = 5$

Lesson 5-1

partition

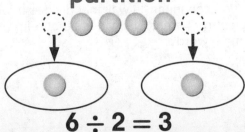

$6 \div 2 = 3$

Lesson 5-4

quotient

$15 \div 3 = 5$

Ideas for Use

- Group 2 or 3 common words. Add a word that is unrelated to the group. Then work with a friend to name the unrelated word.
- Design a crossword puzzle. Use the definition for each word as the clues.

A number that is being divided.

Circle the dividend in $12 \div 4 = \blacksquare$. Then write the quotient.

To separate into equal groups, to find the number of groups, or the number in each group.

How can dividing help you share snacks with friends?

The number by which the dividend is being divided.

Write a division sentence in which the divisor is 5. Circle the divisor.

A number sentence using numbers and the $\div$ sign.

Write an example of a division sentence. Then write the sentence using words.

Operations that undo each other, like multiplication and division.

Use inverse operations to write a multiplication and division sentence.

A group of related facts using the same numbers.

Write the numbers in the fact family shown on this card.

The answer to a division problem.

Write and solve a division problem. Circle the quotient.

To divide or "break up."

Partition can mean "a wall that divides a room into different areas." How does that relate to the math word?

MY Vocabulary Cards

Processes & Practices

Lesson 5-5

related facts

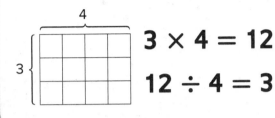

$3 \times 4 = 12$

$12 \div 4 = 3$

Lesson 5-3

repeated subtraction

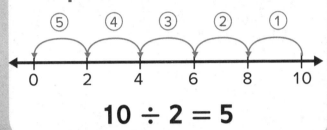

$10 \div 2 = 5$

Ideas for Use

- Write an example for each card. Be sure your examples are different from what is shown on each card.

- Write the name of each lesson on the front of each blank card. Write a few study tips on the back of each card.

Subtraction of the same number over and over again.

Write a sentence comparing repeated subtraction to repeated addition.

Basic facts using the same numbers.

Why is *facts* plural in this vocabulary word?

MY Foldable

FOLDABLES Follow the steps on the back to make your Foldable.

dividend
Definition: The number to be divided.

÷

divisor
Definition: The number the dividend is divided by.

=

quotient
Definition: The answer to a division problem.

 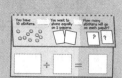 

You have 10 stickers.

↓

You want to share equally on 2 papers.

↓

How many stickers will go on each paper?

 ÷ =

Name _____

Lesson 1
Hands On
Model Division

ESSENTIAL QUESTION
What does division mean?

Division is an operation with two numbers. One number tells you how many items you have. The other tells you how many equal shares, or groups, to form or how many to put in each group.

10 ÷ 5 = 2   *Read ÷ as divided by. 10 divided by 5 = 2.*

To **divide** means to **partition**, or separate a number into equal groups, to find the number of groups, or find the number in each group.

Build It

Find how many in each group.
Divide 12 counters into 3 equal groups.
How many are in each group?

1. Partition one counter at a time into a group until all of the counters are gone.

2. Draw the groups of counters.

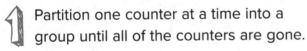

3. Write a **division sentence**, or a number sentence that uses division.

12 counters were divided into

_____ groups.

There are _____ counters in each group.

So, 12 ÷ 3 = _____ in each group.   *SAY: Twelve divided by three equals four.*

Online Content at connectED.mcgraw-hill.com   Lesson 1 Hands On: Model Division   **245**

Try It

Find how many groups. Place 12 counters in groups of 3. How many groups are there?

Make groups of 3 until all the counters are gone. Draw the groups.

Write a division sentence. 12 counters were divided into equal groups of _____ .

There are _____ groups.

12 ÷ 3 = _____ groups.   ← SAY: Twelve divided by three equals four.

Talk About It

1. Explain how you divided 12 counters into equal groups.

   _____
   _____

2. When you divided the counters into groups of 3, how did you find the number of equal groups?

   _____
   _____

3. **Processes &Practices** ③ **Draw a Conclusion** Explain the difference between the way you partitioned the counters in the first activity to the way you partitioned them in the second activity.

   _____
   _____
   _____
   _____

246 Chapter 5 Understand Division

Name ....

Practice It

4. Partition 8 counters one at a time to find the number of counters in each group. Draw the counters.

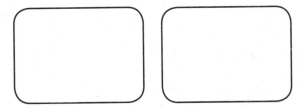

There are _____ counters in each group; 8 ÷ 2 = _____ .

5. Circle equal groups of 5 to find the number of equal groups.

There are _____ equal groups; 15 ÷ _____ = 5.

6. **Algebra** Use counters to find each unknown.

Number of Counters	Number of Equal Groups	Number in Each Group	Division Sentence
9	■	3	9 ÷ ■ = 3
14	2	?	14 ÷ 2 = ?
15	■	5	15 ÷ ■ = 5
6	?	3	6 ÷ ? = 3

7. Choose one division sentence from Exercise 6. Write and solve a real-world problem for that number sentence.

Lesson 1 Hands On: Model Division 247

Apply It

Draw a model to solve. Then write a number sentence.

8. A florist needs to make 5 equal-sized bouquets from 25 flowers. How many flowers will be in each bouquet?

9. **Processes & Practices 4** **Model Math** Mrs. Wilson called the flower shop to place an order for 9 flowers. She wants an equal number of roses, daisies, and tulips. How many of each kind of flower will Mrs. Wilson receive?

10. **Processes & Practices 1** **Make a Plan** Mr. Cutler bought 2 dozen roses to equally arrange in 4 vases. How many roses will he put in each vase? (*Hint:* 1 dozen = 12)

11. **Processes & Practices 2** **Reason** Can 13 counters be partitioned equally into groups of 3? Explain.

Write About It

12. How can I use models to understand division?

248 Chapter 5 Understand Division

Name _____

MY Homework

Lesson 1

Hands On:
Model Division

Homework Helper

Need help? connectED.mcgraw-hill.com

Divide 9 counters into 3 equal groups.
Find how many counters are in each group.

Partition 9 counters, one counter at a time, until all the counters are gone.

9 counters were divided into 3 groups. There are 3 counters in each group.

The division sentence is 9 ÷ 3 = 3.

Practice

1. Partition 6 counters, one at a time, to find the number of counters in each group. Draw the counters.

   _____ counters were divided into 2 groups; 6 ÷ 2 = _____ counters in each group.

2. Circle each group of 4 to find the number of equal groups.

   _____ counters were divided into groups of 4; 16 ÷ 4 = _____ groups.

Problem Solving

Draw a model to solve. Then write a number sentence.

3. Nola has 16 bracelets. She hangs an equal number of bracelets on 2 hooks. How many bracelets are on each hook?

4. **Processes & Practices** **Model Math** Noah rolled 18 large snowballs to make snowmen. He used 3 snowballs for each snowman. How many snowmen did Noah make?

5. There are 8 mittens drying on the heater. Each student has 2 mittens. How many students have mittens drying on the heater?

Vocabulary Check

Draw a line to connect each vocabulary word with its definition.

6. division sentence • placing objects into groups one at a time until there are none left

7. division • a number sentence that shows the number of equal groups and the number in each group

8. partition • groups that have the same quantity of objects or have the same value

9. equal groups • an operation that tells the number of equal groups and the number in each group

250 **Chapter 5** Understand Division

Name _____

Lesson 2
Division as Equal Sharing

ESSENTIAL QUESTION
What does division mean?

One way to **divide** is to find the number in each group. This can be done by equal sharing.

Math in My World

Eat your veggies!

Example 1

Nolan feeds 6 carrots equally to 3 rabbits. How many carrots does each rabbit get?

Draw one carrot at a time next to each rabbit until there are no more carrots.

Write a division sentence to represent the problem. A **division sentence** is a number sentence that uses the operation of division.

_____ carrots equally shared by _____ rabbits gives _____ carrots to each rabbit.

6 ÷ 3 = _____   There are _____ carrots for each rabbit.

You can think of division sentences in two ways.

6 items		6 items
3 equal groups	→ 6 ÷ 3 = 2 ←	2 equal groups
2 items in each group		3 items in each group

Online Content at connectED.mcgraw-hill.com

Lesson 2 251

You can draw an array to help you divide.

Example 2

Fifteen scouts equally shared 3 tents. How many scouts are in each tent? Place one counter (scout) at a time next to each tent until all the counters are gone. Draw a sketch of your counters.

Helpful Hint
When you divide, you share an equal number to all of the groups.

_____ scouts ÷ _____ tents = _____ scouts in each tent

_____ ÷ _____ = _____

There will be _____ scouts in each tent.

Guided Practice

Use counters to find how many are in each group.

1. 10 counters
   2 equal groups

   _____ in each group

   10 ÷ 2 = _____

2. 14 counters
   7 equal groups

   _____ in each group

   14 ÷ 7 = _____

3. 20 counters
   5 equal groups

   _____ in each group

   20 ÷ 5 = _____

Talk MATH
Explain what it means to share equally when dividing.

Name ....

Independent Practice

Use counters to find how many are in each group.

4. 12 counters
2 equal groups

_____ in each group

_____ ÷ _____ = _____

5. 16 counters
4 equal groups

_____ in each group

_____ ÷ _____ = _____

6. 18 counters
6 equal groups

_____ in each group

_____ ÷ _____ = _____

Use counters to find the number of equal groups.

7. 8 counters

_____ equal groups

4 in each group

8 ÷ _____ = 4

8. 21 counters

_____ equal groups

7 in each group

21 ÷ _____ = 7

9. 18 counters

_____ equal groups

9 in each group

18 ÷ _____ = 9

Use counters to draw an array. Write a division sentence.

10. Draw 9 counters in 3 equal rows.

There are _____ in each row.

_____ ÷ _____ = _____

11. Draw 14 counters in 2 equal rows.

There are _____ in each row.

_____ ÷ _____ = _____

Algebra Draw lines to match each division sentence with its correct unknown.

12. 24 ÷ ■ = 3   • 5

13. 30 ÷ 6 = ■   • 7

14. 42 ÷ ■ = 6   • 8

Lesson 2 Division as Equal Sharing   **253**

Problem Solving

Draw a picture to solve. Then write a division sentence.

15. Marla has $25. How many hamster wheels can she buy?

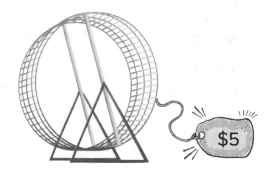

16. A seamstress needs 18 feet of fabric. How many yards of fabric does she need? (*Hint:* 1 yard = 3 feet)

Brain Builders

17. **Processes & Practices** **Plan Your Solution**
There are 6 juice boxes in a package. How many packages need to be bought if 24 juice boxes are needed for a picnic? Write a division sentence with a symbol for the unknown. Then solve.

18. **Processes & Practices** **Model Math** Write a real-world problem that uses the division sentence 12 ÷ 6 = ■. Then find the unknown.

Write a related multiplication sentence to check your answer.

19. **Building on the Essential Question** How is dividing like sharing?

254 **Chapter 5** Understand Division

Name _____

MY Homework

Lesson 2
Division as Equal Sharing

Homework Helper

Need help? connectED.mcgraw-hill.com

There are 16 people on a ride at the fair. They are divided evenly among 4 carts. How many people are in each cart?

Use counters to solve the problem.

1. Start with 16 counters to represent the 16 people.

2. Divide the counters evenly among the carts.

3. 16 riders divided equally among 4 carts is 4 riders per cart.

So, 16 ÷ 4 = 4.

Practice

Use counters to find how many are in each group.

1. 21 counters

   7 equal groups

   _____ in each group

   21 ÷ 7 = _____

2. 16 counters

   2 equal groups

   _____ in each group

   16 ÷ 2 = _____

3. 18 counters

   3 equal groups

   _____ in each group

   18 ÷ 3 = _____

4. 30 counters

   6 equal groups

   _____ in each group

   30 ÷ 6 = _____

Use counters to find the number of equal groups.

5. 24 counters

   _____ equal groups
   3 in each group

   24 ÷ _____ = 3

6. 24 counters

   _____ equal groups
   6 in each group

   24 ÷ _____ = 6

Brain Builders

Draw a picture to solve. Then write a division sentence.

7. Amy and 3 friends want to share 8 apples equally. How many apples will each person get?

8. **Processes & Practices 4** **Model Math** Sarah has 32 crackers. She eats 2 and throws away 2 that she dropped. Sarah puts the rest of the crackers into 4 equal groups. How many crackers are in each group?

Vocabulary Check

Draw an example or write a definition beneath each vocabulary word.

9. array

10. divide

11. division sentence

12. **Test Practice** There are 12 girls and 13 boys in Mr. Copa's class. He divides the students into equal groups of 5. How many students are in each group?

   Ⓐ 5 students
   Ⓑ 10 students
   Ⓒ 15 students
   Ⓓ 20 students

256 **Need more practice?** Download Extra Practice at connectED.mcgraw-hill.com

Name _____

Lesson 3
Relate Division and Subtraction

ESSENTIAL QUESTION
What does division mean?

Math in My World

Example 1

A designer makes 15 dresses in equal numbers of red, blue, and yellow. How many dresses of each color are there? Write a division sentence with a symbol for the unknown. Then solve.

15 ÷ 3 = ■ ← unknown

One Way Use models.

Draw one counter at a time on each dress until all 15 counters are gone.

There are _____ dresses of each color. The unknown is _____.

So, _____ ÷ _____ = _____.

Another Way Use a number line.

You can also divide using **repeated subtraction.** Subtract equal groups of 3 repeatedly until you get to zero.

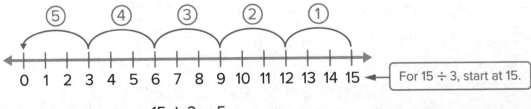

For 15 ÷ 3, start at 15.

15 ÷ 3 = 5

You subtracted groups of three _____ times.

So, 15 ÷ 3 = _____.

Online Content at connectED.mcgraw-hill.com

Lesson 3 257

Example 2

Use repeated subtraction to find 10 ÷ 2. Write a division sentence.

One Way Use a number line.

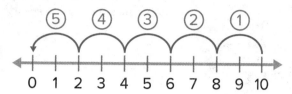

Start at 10. Count back by 2s until you reach 0. How many times did you subtract? _____

So, 10 ÷ 2 = _____.

Another Way Use repeated subtraction.

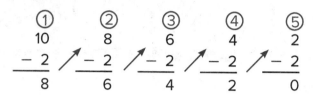

Subtract groups of 2 until you reach 0. How many groups did you subtract? _____

Guided Practice

Algebra Write a division sentence with a symbol for the unknown. Then solve.

1. There are 16 flowers. Each vase has 4 flowers. How many vases are there?

   _____ ÷ 4 = ▪

   There are _____ vases.

2. There are 14 ears. Each dog has 2 ears. How many dogs are there?

   14 ÷ _____ = ▪

   There are _____ dogs.

Use repeated subtraction to divide.

3.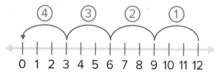

   12 ÷ 3 = _____

4.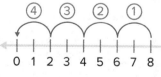

   8 ÷ 2 = _____

Talk MATH

Explain how to use a number line to find 18 ÷ 9.

258 Chapter 5 Understand Division

Name

Independent Practice

Algebra Write a division sentence with a symbol for the unknown. Then solve.

5. There are 16 orange slices. Each orange has 8 slices. How many oranges are there?

6. There are 16 miles. Each trip is 2 miles. How many trips are there?

7. There are 25 marbles, with 5 marbles in each bag. How many bags are there?

8. Four friends will share 12 muffins equally. How many muffins will each friend get?

Use repeated subtraction to divide.

9.

10 ÷ 5 = _____

10.
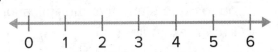
6 ÷ 3 = _____

11.
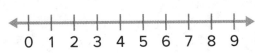
9 ÷ 3 = _____

12.

8 ÷ 4 = _____

13. 12 ÷ 3 = _____
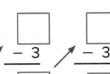

14. 20 ÷ 4 = _____
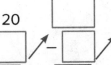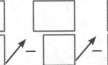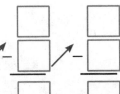

Lesson 3 Relate Division and Subtraction 259

Problem Solving

Chicago's Ferris wheel is 10 stories tall. Each cart can seat up to 6 people for a 7-minute ride.

Write a division sentence with a symbol for the unknown. Then solve.

15. It costs $24 for 4 people to ride the Ferris wheel. How much does each ticket cost?

_____

16. **Processes &Practices 2** **Use Symbols** If 30 students from a class wanted to ride, how many carts would they need?

_____

Brain Builders

17. **Processes &Practices 2** **Reason** How can knowing that multiplication is repeated addition and division is repeated subtraction help you to understand that multiplication and division are related? Give an example.

_____
_____
_____

18. **Building on the Essential Question** How is division related to subtraction?

_____
_____

Explain how you could use counters to show the relationship.

_____
_____
_____

260 Chapter 5 Understand Division

Name

Lesson 3

Relate Division and Subtraction

Homework Helper

Need help? connectED.mcgraw-hill.com

Perry divides 9 berries evenly among 3 fruit cups. How many berries does Perry put in each cup? Write a division sentence with a symbol for the unknown. Solve.

9 ÷ 3 = ■ ← unknown

One Way Use a number line.

Subtract equal groups of 3 until you reach 0. There are 3 groups.

So, 9 ÷ 3 = 3.

Another Way Use repeated subtraction.

```
  ①      ②      ③
  9       6       3
 −3      −3      −3
 ───     ───     ───
  6       3       0
```

Keep subtracting 3 until you reach 0. You subtracted 3 groups.

So, 9 ÷ 3 = 3.

Practice

Use repeated subtraction to divide.

1.

```
├─┼─┼─┼─┼─┼─┼─┼─┼─┼─┼─┼─┼─┼─┤
0 1 2 3 4 5 6 7 8 9 10 11 12 13 14
```

14 ÷ 2 = _____

2.

```
├─┼─┼─┼─┼─┼─┼─┼─┼─┼─┼─┼─┤
0 1 2 3 4 5 6 7 8 9 10 11 12
```

12 ÷ 6 = _____

3. 28 ÷ 7 = _____

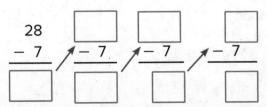

4. 30 ÷ 10 = _____

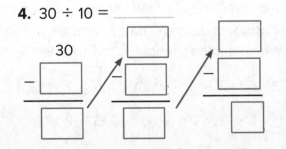

Lesson 3 My Homework 261

Problem Solving

Algebra Write a division sentence with a symbol for the unknown. Then solve.

5. There are 24 juice bottles, with 6 bottles in each package. How many packages of juice bottles are there?

_____

6. A mechanic equally divides 4 hours of his time to repair 8 cars. How many cars does he repair in one hour?

_____

Brain Builders

7. On Monday, Helen's math teacher gave the class 45 problems to finish by Friday. Helen will do the same number of problems each day. How many problems will she do on Thursday?

_____

Vocabulary Check

Choose the correct word(s) to complete each sentence.

number line   repeated subtraction

8. Skip count backward on a _____ to divide.

9. Use _____ to subtract equal groups repeatedly until you reach zero.

10. **Test Practice** Sal made 6 cups of oatmeal for himself and his 2 brothers. Each received an equal number of cups of oatmeal. Which number sentence represents this problem?

   Ⓐ $6 - 2 = 4$ Ⓒ $6 \times 2 = 12$

   Ⓑ $6 \div 3 = 2$ Ⓓ $6 - 3 = 3$

Check My Progress

Vocabulary Check

Use the word bank to label each definition.

| array | division | division sentence |
| equal groups | repeated subtraction | |

1.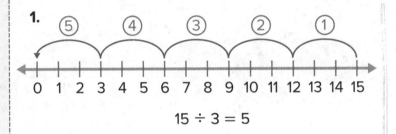
   $15 \div 3 = 5$

   _____

2.
   $10 \div 5 = 2$

   _____

3.

   _____

4.

   _____

5. _____ is an operation where one number tells you how many things you have and the other tells you how many equal groups to form or how many to put in each group.

Concept Check

Write a division sentence and divide to find how many are in each group.

6. 12 counters
   3 equal groups

   _____ ÷ _____ = _____

   There are _____ in each group.

7. 15 counters
   5 equal groups

   _____ ÷ _____ = _____

   There are _____ in each group.

Use repeated subtraction to divide.

8.

   12 ÷ 4 = _____

9.

   16 ÷ 8 = _____

Brain Builders

Algebra Write a division sentence with a symbol for the unknown. Then solve.

10. Coach Shelton divided 12 girls and 6 boys into 3 equal-sized teams. How many players are on each team?

    _____

11. Chang has 5 turtles and 10 frogs in his pond. If he catches 3 animals a day, how many days will it take him to catch all of the animals?

    _____

12. **Test Practice** Kayla spends $20 to buy 2 vanilla and 2 peppermint candles. Each candle is the same price. What is the cost of one candle?

    Ⓐ $4   Ⓒ $16
    Ⓑ $5   Ⓓ $24

264 **Chapter 5** Understand Division

Name _____

Lesson 4
Hands On
Relate Division and Multiplication

ESSENTIAL QUESTION
What does division mean?

Division and multiplication are related operations.

Build It

Find 21 ÷ 3.

1. Model 21 counters divided into 3 equal groups. Draw the model. How many counters are in each group?

   _____ counters

2. Write a division sentence.

   number in all ÷ number of groups = number in each group

   ☐ ÷ ☐ = ☐

   The **dividend** is the number to be divided.
   The **divisor** is the number by which the dividend is divided.
   The answer is the **quotient.**

3. Write a multiplication sentence.

   number of groups × number in each group = number in all

   ☐ × ☐ = ☐

Online Content at connectED.mcgraw-hill.com

265

Try It

Find 20 ÷ 4.

 Model 20 connecting cubes divided into 4 equal rows. Draw the model. How many cubes are in each row?

_____ cubes

 Write a division sentence.

dividend		divisor		quotient
☐	÷	☐	=	☐

 Write a multiplication sentence.

factor		factor		product
☐	×	☐	=	☐

Talk About It

1. Explain how you used models to show 21 ÷ 3.

2. **Processes & Practices** **Explain to a Friend** Explain how the array shows that 21 ÷ 3 = 7 is related to 3 × 7 = 21.

3. **Processes & Practices** **Look for a Pattern** What pattern do you notice between the number sentences in the two activities?

4. How can multiplication facts be used to divide?

Name _____

Practice It

Write a related division and multiplication sentence for each.

5.

6.

_____

_____

Use connecting cubes to solve. Draw your model. Write a division sentence.

7. Model 10 connecting cubes divided into 2 equal rows. How many cubes are in each row?

8. Model 6 connecting cubes divided into 3 equal rows. How many cubes are in each row?

_____

_____

Algebra Use counters to model each problem. Find the unknown. Then write a related multiplication sentence.

9. 12 ÷ 6 = ■

10. 21 ÷ 7 = ■

11. 25 ÷ 5 = ■

The unknown is ____.

The unknown is ____.

The unknown is ____.

_____

_____

_____

Lesson 4 Hands On: Relate Division and Multiplication   267

Apply It

Draw a model to solve. Then write a division sentence.

12. A scientist organizes 14 bugs into 2 equal rows. How many bugs are in each row?

13. A teacher divides her 24 students into 4 equal activity groups. How many students are in each group?

14. Arianna divides 25 gold stars between herself and 4 friends. How many stars does each friend receive?

15. Mark checked out 12 books from the library. He reads an equal amount of books each week for 4 weeks. How many books does he read each week?

16. **Processes & Practices 1** **Make a Plan** Eli had 20 lemons. He used an equal number in each of 3 pitchers of lemonade. He has 2 lemons left. How many lemons did Eli use to make one pitcher of lemonade?

Write About It

17. How are arrays used in both multiplication and division?

Name ....

Lesson 4
Hands On: Relate Division and Multiplication

Homework Helper

Need help? connectED.mcgraw-hill.com

Find $32 \div 4$.

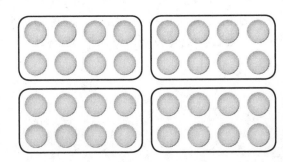

1. Model 32 counters divided into 4 equal groups. There are 8 counters in each group.

2. Write a division sentence.

$$32 \div 4 = 8$$

dividend divisor quotient

3. Write a multiplication sentence.

$$4 \times 8 = 32$$

number of groups — number in each group — number in all

So, $32 \div 4 = 8$.

Practice

Write a related division and multiplication sentence for each.

1.

2.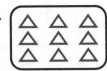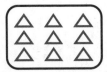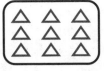

Lesson 4 My Homework 269

Problem Solving

Draw a model to solve. Then write a division sentence.

3. 42 students need to divide equally into 7 vans going to the museum. How many students will be in each van?

_____

4. **Processes &Practices** **Model Math** Carla is giving out pencils to 30 students. The students are divided equally among 6 tables. How many pencils will Carla leave at each table?

_____

5. Mr. Rina has 7 glass figures. He will use 1 box to mail each glass figure to a customer. How many boxes does Mr. Rina need?

_____

Vocabulary Check

Use the vocabulary to label each number in the division sentence.

quotient   divisor   dividend

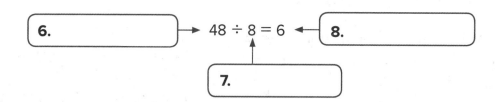

270 Chapter 5 Understand Division

Name _____

Lesson 5
Inverse Operations

ESSENTIAL QUESTION
What does division mean?

You have learned how division and multiplication are related. Operations that are related are **inverse operations** because they undo each other.

Math in My World

Example 1

A baker has made a tray of fresh muffins. Use the array to write a related multiplication and division sentence to find the unknown. How many muffins are there in all?

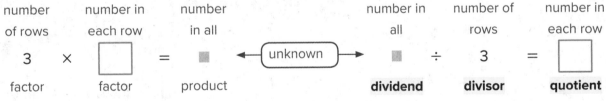

Multiplication

number of rows	number in each row	number in all
3	× ☐	= ■
factor	factor	product

← unknown →

Division

number in all	number of rows	number in each row
■	÷ 3	= ☐
dividend	**divisor**	**quotient**

The unknown is _____ .

So, there are _____ muffins in all.

The multiplication sentence multiplies 3 by _____ to get 12. The division sentence undoes the multiplication by dividing 12 by 3 to get _____ .

Online Content at connectED.mcgraw-hill.com

A group of **related facts** using the same numbers is a **fact family**.
Each fact family follows a pattern by using the same numbers.

Fact Family 3, 4 and 12

$3 \times 4 = 12$
$4 \times 3 = 12$
$12 \div 3 = 4$
$12 \div 4 = 3$

Fact Family 7 and 49

$7 \times 7 = 49$
$49 \div 7 = 7$

Example 2

Complete the fact family for the numbers 3, 6, and 18.

$3 \times 6 = \square$ $18 \div \square = 6$

$\square \times 3 = 18$ $18 \div 6 = \square$

The pattern shows that 3, 6, and 18 are used in each number sentence.

Guided Practice

Use the arrays to find each unknown.

1. $\blacksquare \times 5 = 15$

   $? \div 3 = 5$

   $\blacksquare = $ _____

   $? = $ _____

2. $4 \times ? = 24$

   $24 \div \blacksquare = 6$

   $? = $ _____

   $\blacksquare = $ _____

3. Write the fact family for 2, 6, and 12.

   $\square \times 6 = \square$ $12 \div \square = \square$

   $\square \times \square = 12$ $\square \div \square = 2$

Why are the product and the dividend the same in $3 \times 7 = 21$ and $21 \div 3 = 7$?

Name

Independent Practice

Algebra Use the arrays and inverse operations to find each unknown.

4. ■ × 2 = 8

   ? ÷ 4 = 2

   ■ = _____

   ? = _____

5. 2 × ? = 4

   4 ÷ ? = 2

   ? = _____

6. ? × 2 = 14

   ■ ÷ 2 = 7

   ■ = _____

   ? = _____

7. 4 × ■ = 20

   20 ÷ ■ = 4

   ■ = _____

Write the fact family for each set of numbers.

8. 2, 3, 6

9. 2, 7, 14

10. 4, 8, 32

11. 4, 3, 12

Write the set of numbers for each fact family.

12. 5 × 9 = 45
    9 × 5 = 45
    45 ÷ 9 = 5
    45 ÷ 5 = 9

13. 7 × 4 = 28
    4 × 7 = 28
    28 ÷ 7 = 4
    28 ÷ 4 = 7

14. 3 × 3 = 9
    9 ÷ 3 = 3

Lesson 5 Inverse Operations 273

Problem Solving

Write a division sentence to solve.

15. All 5 members of the Malone family went to the movies. Their tickets cost a total of $30. How much was each ticket?

16. The petting zoo has 21 animals. There are an equal number of goats, ponies, and cows. How many of each animal are there?

Brain Builders

17. **Processes & Practices 5** **Use Math Tools** Mr. Thomas travels a total of 20 miles each week to and from work. If he works 5 days a week, how many miles does Mr. Thomas live from where he works?

18. **Processes & Practices 3** **Draw a Conclusion** Look back to Exercise 14 on the previous page. Why are there only 2 numbers in the fact family instead of 3 numbers? Give another example of a fact family with only 2 numbers.

19. **Building on the Essential Question** How can I use multiplication facts to remember division facts? Give an example.

MY Homework

Lesson 5

Inverse Operations

Homework Helper

Need help? connectED.mcgraw-hill.com

The array represents 27 children lined up in 3 rows. Use the array to find each unknown.

9 × ■ = 27
? ÷ 3 = 9
■ = 3
? = 27

You know 3 rows of 9 = 27. So, 9 rows of 3 = 27 and 27 ÷ 3 = 9.

Practice

Algebra Use the array to find each unknown.

1. ■ × 4 = 20
   ? ÷ 5 = 4
   ■ = _____
   ? = _____

2. 4 × ■ = 16
   ? ÷ 4 = 4
   ■ = _____
   ? = _____

3. 7 × ■ = 21
   ? ÷ 7 = 3
   ■ = _____
   ? = _____

4. 2 × ■ = 12
   ? ÷ 2 = 6
   ■ = _____
   ? = _____

Write the fact family for each set of numbers.

5. 5, 8, 40

6. 6, 7, 42

Write the set of numbers for each fact family.

7. $4 \times 9 = 36$ $36 \div 4 = 9$
   $9 \times 4 = 36$ $36 \div 9 = 4$

8. $2 \times 8 = 16$ $16 \div 2 = 8$
   $8 \times 2 = 16$ $16 \div 8 = 2$

Brain Builders

9. **Processes & Practices** **Model Math** Tia finds $8 in her pocket. She uses it and the $27 she has saved to buy socks for the family. If each pair of socks costs $5, how many pairs can she buy? Write a division sentence to solve.

Vocabulary Check

Draw a line to connect each vocabulary word with its definition.

10. dividend   • the number being divided
11. divisor   • a group of related facts that use the same numbers
12. fact family   • the answer to a division problem
13. inverse operations • the number by which a dividend is divided
14. quotient   • operations that undo each other

15. **Test Practice** Which pair shows inverse operations?

   Ⓐ $2 \times 2 = 4; 4 \div 2 = 2$
   Ⓑ $2 \times 2 = 4; 4 - 2 = 2$
   Ⓒ $2 \times 2 = 4; 8 \div 4 = 2$
   Ⓓ $2 \times 2 = 4; 4 \div 4 = 1$

Name _____

Lesson 6
Problem-Solving Investigation
STRATEGY: Use Models

ESSENTIAL QUESTION
What does division mean?

Learn the Strategy

Mia has 18 items that need to be split evenly among 3 welcome baskets. How many items will Mia put in each basket?

1 Understand
What facts do you know?

_____ items need to be split

evenly among _____ baskets.

What do you need to find?

the number of _____

2 Plan
I will make a model to find _____.

3 Solve

I will use counters to model the problem

by placing _____ counter at a time in each group.

The model shows that 18 ÷ 3 = _____.

So, Mia will fill each basket with _____ items.

4 Check
Does your answer make sense? Explain.

_____

Practice the Strategy

A veterinarian helped 20 pets from Monday to Friday. She helped an equal number of pets each day. How many pets did she help each day?

1 Understand

What facts do you know?

_____

_____

What do you need to find?

_____

2 Plan

_____

3 Solve

4 Check

Does your answer make sense? Explain.

_____

_____

Name

Apply the Strategy

Solve each problem by using a model.

1. **Processes & Practices 5** **Use Math Tools** Jill has 27 blocks. She wants to divide them equally into the bowls shown below. How many blocks will be in each bowl?

2. The owner of an apartment building needs to fix 16 locks in four of his apartments. Each apartment has the same number of locks that need to be fixed. How many locks in each apartment need to be fixed?

Brain Builders

3. A baker used a dozen eggs to make 1 batch of cupcakes and 2 cakes. The recipe called for each batch of cupcakes and cake to have the same number of eggs. How many eggs did the baker use in each batch? (*Hint:* 1 dozen = 12)

4. There are 13 girls and 11 boys that want to play a game. They need to make 4 teams. How many players will be on each team if each team needs an equal number of players?

Review the Strategies

Use any strategy to solve each problem.
- Determine reasonable answers.
- Use an estimate or exact answer.
- Use models.

5. Processes &Practices 2 Use Number Sense Sarah needs 15 pieces of chalk for a project. Each box contains 3 pieces of chalk. How many boxes of chalk will she need to buy?

6. Brooke volunteers to read with young children 5 nights a month. She spends 2 hours each visit. This month, she volunteered one extra night. How many hours did she read with the children this month?

7. Processes &Practices 4 Model Math A chef will make pizzas. He has broccoli, peppers, onions, pepperoni, and sausage. How many types of pizzas can be made with one type of vegetable and one type of meat? Name the combinations.

8. A scientist estimates that a brown bear weighs 700 pounds. It actually weighs 634 pounds. How much more is the estimate than the actual weight?

MY Homework

Lesson 6

Problem Solving:
Use Models

Homework Helper

Need help? connectED.mcgraw-hill.com

Lucy needs 7 craft sticks to make a puzzle. She has 28 craft sticks. How many puzzles can Lucy make? Use a model to solve.

1 Understand
Lucy has 28 craft sticks. She needs 7 sticks to make a puzzle. Find how many puzzles Lucy can make.

2 Plan
Divide 28 craft sticks into equal groups of 7.

3 Solve
There are 4 equal groups of 7 craft sticks.
The model shows that $28 \div 7 = 4$.
So, Lucy can make 4 puzzles.

4 Check
Use multiplication to check. $4 \times 7 = 28$
So, the answer is correct.

Problem Solving

1. Brandon spent $20 on school supplies. He bought five different items that each cost the same amount. How much did each item cost? Use a model to solve.

   Each item cost _____ .

Solve each problem by using a model.

2. **Processes &Practices** 5 **Use Math Tools** Alice planted 6 tomato plants, 4 bean plants, and 2 pepper plants. Each row had 6 plants. How many rows did Alice plant?

3. At the circus, there are 18 clowns. The clowns drive around in little cars. If there are 3 clowns in each car, how many cars are there?

Brain Builders

4. Mr. and Mrs. Carson took Sarah, Brent, and Joanie to see a movie. The Carsons spent $15 on snacks. They paid $50 in all. How much did each ticket cost?

5. Mrs. Glover had 14 rare coins. Mr. Glover had 11 rare coins. They divided them evenly among 5 grandchildren. How many coins did each grandchild get?

6. A singer performed 5 fast and 4 slow songs at a recital. She had 3 weeks to practice. How many songs did she practice each week if she practiced an equal number of songs each week?

Review

Chapter 5
Understand Division

Vocabulary Check

Use the word bank to complete each sentence.

| array | divide | fact family | inverse operations |
| partition | related facts | repeated subtraction | |

1. _____ are a set of basic facts using the same three numbers.

2. An arrangement of objects into equal rows and equal columns is a(n) _____.

3. A way to divide by sharing one object at a time until all the objects are gone is to _____.

4. To _____ means to separate a number into equal groups, to find the number of groups, or find the number in each group.

5. _____ is a way to subtract the same number over and over again until you reach 0.

6. Operations that are related are _____ because they undo each other.

7. 3 × 5 = 15, 5 × 3 = 15, 15 ÷ 5 = 3, and 15 ÷ 3 = 5 are the facts in the 3, 5, 15 _____.

8. Write a **division sentence** in the space below. Label the **dividend**, **divisor**, and **quotient**.

My Chapter Review 283

Concept Check

Use counters to find how many are in each group.

9. 14 counters

   2 equal groups

   _____ ÷ _____ = _____

   _____ in each group

10. 25 counters

    5 equal groups

    _____ ÷ _____ = _____

    _____ in each group

Use repeated subtraction to divide.

11.

    $12 \div 6 =$ _____

12.

    $20 \div 4 =$ _____

Write a related division and multiplication sentence for each.

13.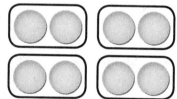

14.

Write the fact family for each set of numbers.

15. 4, 7, 28

16. 3, 9, 27

284 Chapter 5 Understand Division

Problem Solving

17. Brandon's dentist gave him 12 toothbrushes. Brandon wants to share them equally among himself and his 2 friends. How many toothbrushes will each person get? Write a division sentence.

18. A teacher has 24 pencils. She keeps 4 and shares the others equally with 5 students. How many pencils did each student get?

Brain Builders

19. Circle the number sentence that does not belong. Explain. Then write the missing number sentence.

$3 \times 6 = 18$ $18 \div 2 = 9$

$18 \div 6 = 3$ $6 \times 3 = 18$

20. Test Practice Harper saved $20 from mowing lawns and $10 from walking dogs April through September. She saved an equal amount each month. How much money did Harper save each month?

Ⓐ $5   Ⓒ $8

Ⓑ $6   Ⓓ $10

Reflect

Chapter 5 Answering the ESSENTIAL QUESTION

Use what you learned about division to complete the graphic organizer.

Real-World Problem

Vocabulary

Draw a Model

ESSENTIAL QUESTION What does division mean?

Write a Number Sentence

Reflect on the ESSENTIAL QUESTION. Write your answer below.

Name _____ Date _____

Score _____

Performance Task

Academic Challenge

The school is having tryouts for students to select a team for the Academic Challenge. The Challenge asks academic questions of the students and is shown on television. They will compete against other schools.

Show all your work to receive full credit.

Part A

At one tryout, each student is given 3 pieces of paper to fill out. If 27 pieces of paper are given to students, how many students are at the tryout? Draw a model to help you find the answer. Write the division sentence shown by the model.

Part B

At another tryout, the teachers divide the students into groups of 4 students each. If there are 28 students in all at the tryout, how many groups are there? Draw a model to support your answer.

Online Content at connectED.mcgraw-hill.com

Part C

For some of the practices for the Academic Challenge, the teachers separate the 11-year olds and the 12-year olds into equal groups. They want to have five equal groups of 11-year olds and seven equal groups of 12-year olds. If there are twenty 11-year olds and forty-two 12-year olds at this practice, how many 11-year olds are in each group? How many 12-year olds are in each group? Show your work.

Part D

Five students are selected to try out for the captain of the team. Each student is given an equal number of questions to answer. If 40 questions are asked in all, how many questions does each student answer? Write a division sentence to support your answer.

Part E

At the final practice, the teachers brought pretzels for the students. There were 63 pretzels in all. The 9 students on the team divided the pretzels equally. How many pretzels did each player get? Use repeated subtraction to support your answer.

Chapter 6
Multiplication and Division Patterns

ESSENTIAL QUESTION

What is the importance of patterns in learning multiplication and division?

Let's Collect!

Watch a video!

Name ....

MY Chapter Project

Clothing Drive

1. As a class, brainstorm local charities that need clothing donations. Choose two or more charities the class will support and list them below.

2. Decorate the donation boxes and label each of the boxes with the name of the charity. Set goals for the number of items the class wants to collect for each charity and write them below.

3. Write a multiplication sentence to show the total donation goal. Explain what the factors and product represent.

Name _____

Am I Ready?

Multiply.

1. $6 \times 4 =$ _____
2. $1 \times 5 =$ _____
3. $7 \times 2 =$ _____

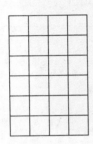 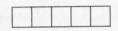 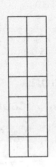

Draw an array for each. Multiply.

4. $4 \times 5 =$ _____
5. $1 \times 6 =$ _____
6. $2 \times 9 =$ _____

Identify a pattern. Then find the missing numbers.

7. _____, _____, 30, 25, 20, 15

    The pattern is _____.

8. _____, _____, 16, 14, 12, 10

    The pattern is _____.

9. Louis has 2 quarters. Yellow whistles cost 5¢ each. Louis wants to buy 8 whistles. Does he have enough money? Explain.

10. Nine trees lined each side of a street. Some trees were cut down leaving a total of 7 trees. How many trees were cut down?

How Did I Do? Shade the boxes to show the problems you answered correctly.

| 1 | 2 | 3 | 4 | 5 | 6 | 7 | 8 | 9 | 10 |

Online Content at connectED.mcgraw-hill.com

Name ....

MY Math Words

Review Vocabulary

bar diagram   factor   partition   product

Making Connections
Choose one review vocabulary word. Use the graphic organizer below to write about and draw examples of the word.

My Description

When I Use This in Math

My Example

My Non-Example

290 Chapter 6 Multiplication and Division Patterns

MY Vocabulary Cards

Lesson 6-8

multiple

multiples of 10:

0, 10, 20, 30, 40

Ideas for Use

- Write a tally mark each time you read the word in this chapter or use the word in your writing. Challenge yourself to making at least ten tally marks for the word.

- Use the blank cards to write review vocabulary cards. Choose review words from this chapter, such as *factor*, *product*, or *partition*.

A multiple of a number is the product of that number and any other number.

How can the term *multiplication* help you remember what a multiple is?

MY Foldable

FOLDABLES Follow the steps on the back to make your Foldable.

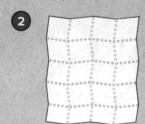
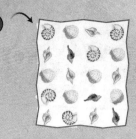

Name _____

Lesson 1
Patterns in the Multiplication Table

ESSENTIAL QUESTION
What is the importance of patterns in learning multiplication and division?

Patterns in the multiplication table can help you remember products and find unknown factors.

Math in My World

Example 1

Enrique noticed he could find the product of two factors in the multiplication table. What is the product of 2 × 3?

The **black numbers** in the table are products. The column and row of **blue numbers** are factors.

1. Look at the two circled factors. Follow the numbers across and down until they meet. This is the product. Complete the number sentence.

   factors
   2 × 3 = _____ ← product

2. Draw a triangle around the product in the multiplication table that has the same factors. Follow left and above to its factors. Draw a triangle around each factor. Complete the number sentence.

   _____ × _____ = 6

   The two number sentences are examples of the

   _____ Property of Multiplication.

×	0	1	2	3	4	5	6	7	8	9	10
0	0	0	0	0	0	0	0	0	0	0	0
1	0	1	2	3	4	5	6	7	8	9	10
2	0	2	4	6	8	10	12	14	16	18	20
3	0	3	6	9	12	15	18	21	24	27	30
4	0	4	8	12	16	20	24	28	32	36	40
5	0	5	10	15	20	25	30	35	40	45	50
6	0	6	12	18	24	30	36	42	48	54	60
7	0	7	14	21	28	35	42	49	56	63	70
8	0	8	16	24	32	40	48	56	64	72	80
9	0	9	18	27	36	45	54	63	72	81	90
10	0	10	20	30	40	50	60	70	80	90	100

Online Content at connectED.mcgraw-hill.com

Lesson 1 295

Example 2

Enrique found a pattern when he multiplied 4 by any factor.

Use a yellow crayon to finish Enrique's pattern. Write the numbers.

0, 4, 8, 12, _____, _____, _____,

_____, _____, _____.

Circle whether the product of 4 and any number is even or odd.

    even   odd

The product of 4 and 5 is 20. Write this product as the sum of two equal numbers.

_____ + _____ = 20

×	0	1	2	3	4	5	6	7	8	9	10
0	0	0	0	0	0	0	0	0	0	0	0
1	0	1	2	3	4	5	6	7	8	9	10
2	0	2	4	6	8	10	12	14	16	18	20
3	0	3	6	9	12	15	18	21	24	27	30
4	0	4	8	12	16	20	24	28	32	36	40
5	0	5	10	15	20	25	30	35	40	45	50
6	0	6	12	18	24	30	36	42	48	54	60
7	0	7	14	21	28	35	42	49	56	63	70
8	0	8	16	24	32	40	48	56	64	72	80
9	0	9	18	27	36	45	54	63	72	81	90
10	0	10	20	30	40	50	60	70	80	90	100

Example 3

Use a blue crayon to color the products with a factor of 3. What do you notice about these products?

The list of products with a factor of _____ increase by _____. It is as if you are counting by 3s.

Guided Practice

1. Use an orange crayon to color the products with a factor of 5. What do you notice about the products in this row and column?

   The products with a factor of _____ end in _____ or _____.

2. Use a purple crayon to color the products with a factor of 10. What do you notice about the products in this row and column?

   The products with a factor of _____ end in _____.

×	0	1	2	3	4	5	6	7	8	9	10
0	0	0	0	0	0	0	0	0	0	0	0
1	0	1	2	3	4	5	6	7	8	9	10
2	0	2	4	6	8	10	12	14	16	18	20
3	0	3	6	9	12	15	18	21	24	27	30
4	0	4	8	12	16	20	24	28	32	36	40
5	0	5	10	15	20	25	30	35	40	45	50
6	0	6	12	18	24	30	36	42	48	54	60
7	0	7	14	21	28	35	42	49	56	63	70
8	0	8	16	24	32	40	48	56	64	72	80
9	0	9	18	27	36	45	54	63	72	81	90
10	0	10	20	30	40	50	60	70	80	90	100

Chapter 6 Multiplication and Division Patterns

Name _____

Independent Practice

3. Shade a row of numbers **blue** that show the products with a factor of 2. What do you notice about the products in this row?

   The 2s products end in

   _____, _____, _____,

   _____, or _____.

   Are all the products in this row even or odd?

   _____

×	0	1	2	3	4	5	6	7	8	9	10
0	0	0	0	0	0	0	0	0	0	0	0
1	0	1	2	3	4	5	6	7	8	9	10
2	0	2	4	6	8	10	12	14	16	18	20
3	0	3	6	9	12	15	18	21	24	27	30
4	0	4	8	12	16	20	24	28	32	36	40
5	0	5	10	15	20	25	30	35	40	45	50
6	0	6	12	18	24	30	36	42	48	54	60
7	0	7	14	21	28	35	42	49	56	63	70
8	0	8	16	24	32	40	48	56	64	72	80
9	0	9	18	27	36	45	54	63	72	81	90
10	0	10	20	30	40	50	60	70	80	90	100

4. Shade a column of numbers **green** that show the products with a factor of 3. Describe the pattern of even and odd products.

   _____

5. Shade a row of numbers **yellow** that show the products with a factor of 1. What do you notice about this row?

   _____

6. Look at the product shaded **gray**. Circle the two factors that make this product. Complete the number sentence.

   4 × _____ = 36

   Draw a triangle around the product that has the same factors. Draw a triangle around each factor. Complete the number sentence.

   9 × _____ = 36

   The two number sentences show the _____ Property of Multiplication.

Lesson 1 Patterns in the Multiplication Table 297

Problem Solving

7. Layne packed 3 toy cars in each of 4 cases. Circle the factors and shade the product to find how many toy cars Layne packed.

_____

Joey packed 4 toy cars in each of 3 cases. Circle the other two factors and shade the product to find how many toy cars Joey packed.

×	0	1	2	3	4	5	6	7	8	9	10
0	0	0	0	0	0	0	0	0	0	0	0
1	0	1	2	3	4	5	6	7	8	9	10
2	0	2	4	6	8	10	12	14	16	18	20
3	0	3	6	9	12	15	18	21	24	27	30
4	0	4	8	12	16	20	24	28	32	36	40
5	0	5	10	15	20	25	30	35	40	45	50

_____

8. Write the two number sentences that show the ways each boy packed the toy cars in Exercise 7.

_____

Which property is this an example of?

_____ Property of _____.

Brain Builders

9. **Processes &Practices 7** **Identify Structure** Write a real-world problem for which you can use the multiplication table and the Commutative Property of Multiplication to solve. Then solve.

_____

_____

10. **Building on the Essential Question** How can a multiplication table help you multiply?

_____

_____

Name: _____

Lesson 1

Patterns in the Multiplication Table

Homework Helper

Need help? connectED.mcgraw-hill.com

Find the product of 3 × 4.

 Find 3 in the far left column.

② Find 4 in the row along the top.

③ Follow the numbers across and down until they meet. This is the product.

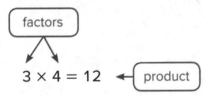

×	0	1	2	3	4	5	6	7	8	9	10
0	0	0	0	0	0	0	0	0	0	0	0
1	0	1	2	3	4	5	6	7	8	9	10
2	0	2	4	6	8	10	12	14	16	18	20
3	0	3	6	9	12	15	18	21	24	27	30
4	0	4	8	12	16	20	24	28	32	36	40
5	0	5	10	15	20	25	30	35	40	45	50
6	0	6	12	18	24	30	36	42	48	54	60
7	0	7	14	21	28	35	42	49	56	63	70
8	0	8	16	24	32	40	48	56	64	72	80
9	0	9	18	27	36	45	54	63	72	81	90
10	0	10	20	30	40	50	60	70	80	90	100

The Commutative Property tells you that you can change the order of the factors without changing the product.

factors
4 × 3 = 12 ← product

Practice

1. Look at the products with a factor of 5. What pattern do you see?

   The products with a factor of 5 end in _____ or _____.

2. Look at the products with a factor of 0. What do you notice?

   The products with a factor of 0 end in _____.

Lesson 1 My Homework 299

3. Find 10 × 5. Circle the factors and the product. Write the product.

_____

4. Shade a row of numbers yellow to show the products with a factor of 10. What do you notice about this row?

The products with a factor of 10 end in _____.

×	0	1	2	3	4	5	6	7	8	9	10
0	0	0	0	0	0	0	0	0	0	0	0
1	0	1	2	3	4	5	6	7	8	9	10
2	0	2	4	6	8	10	12	14	16	18	20
3	0	3	6	9	12	15	18	21	24	27	30
4	0	4	8	12	16	20	24	28	32	36	40
5	0	5	10	15	20	25	30	35	40	45	50
6	0	6	12	18	24	30	36	42	48	54	60
7	0	7	14	21	28	35	42	49	56	63	70
8	0	8	16	24	32	40	48	56	64	72	80
9	0	9	18	27	36	45	54	63	72	81	90
10	0	10	20	30	40	50	60	70	80	90	100

Brain Builders

5. **Processes &Practices** **Model Math** Mason has 1 notebook for science and 1 notebook for reading. He put 9 stickers on each notebook. How many stickers did Mason use in all? Write two multiplication sentences.

_____

Vocabulary Check

6. Label each with the correct word.

factors

product   4 × 2 = 8

7. **Test Practice** Carol wrote two multiplication sentences using the Commutative Property. Which two sentences did she write?

Ⓐ 4 × (2 × 3) = 24 and (4 × 2) × 2 = 16

Ⓑ 2 × (4 × 5) = 40 and 2 × (5 × 4) = 40

Ⓒ 4 × 6 = 24 and 24 = 4 × 6

Ⓓ 8 × 1 = 8; 8 × 0 = 0

Name _____

Lesson 2
Multiply by 2

ESSENTIAL QUESTION
What is the importance of patterns in learning multiplication and division?

Math in My World

Group Project!

Example 1

The students in an art class are working on a project. How many students are there in the art class if there are **8 groups of 2**?

Find 8 groups of 2.

Write 8 groups of 2 as 8 × 2.

One Way Use an array.

Draw an array with 8 rows and 2 columns.

Another Way Draw a picture.
Draw 8 equal groups of 2.

Write an addition sentence and multiplication sentence.

___ + ___ + ___ + ___ + ___ + ___ + ___ + ___ = ___   ___ × ___ = ___

So, 8 × 2 = _____ . There are _____ students in the art class.

$$\begin{array}{r} 8 \\ \times\, 2 \\ \hline 16 \end{array}$$

You can write it vertically, also.

No matter which way you write a multiplication fact, you still read it the same way.

Online Content at connectED.mcgraw-hill.com

Lesson 2 **301**

Example 2

Seth rides his bike to the park Mondays, Wednesdays, and Fridays. It is 2 miles round-trip. How many miles does he ride for the three days? Write a multiplication sentence with a symbol for the unknown. Then, use a bar diagram to solve.

3 × 2 = ■ ← unknown

1 Model 2 miles a day as one part.
1 part = 2 miles

2 Since he rode the same amount for 3 days, model a total of 3 parts.

|2 miles|
|-- 1 day --|

? miles
| 2 miles | 2 miles | 2 miles |
-------- 3 days --------

3 Write a multiplication sentence.   _____ days × _____ miles a day = _____ miles

So, 3 × 2 = _____.

Seth rode _____ miles for the 3 days. The unknown is _____.

Guided Practice

Write an addition sentence and a multiplication sentence for each.

1.

   4 groups of 2 is _____.

   2 + 2 + _____ + _____ = _____

   4 × _____ = _____

2.

   3 groups of 2 is _____.

   2 + 2 + _____ = _____

   3 × _____ = _____

Describe two strategies you can use to remember the multiplication facts for 2.

Name _____

Independent Practice

Write an addition sentence and a multiplication sentence for each.

3.

    2 groups of 2 is _____ .

    2 + _____ = _____

    2 × _____ = _____

$$\begin{array}{r} 2 \\ \times\, 2 \\ \hline \Box \end{array}$$

4.

    6 groups of 2 is _____ .

    2 + 2 + 2 + _____ + _____ + _____ = _____

    6 × _____ = _____

$$\begin{array}{r} 6 \\ \times\, 2 \\ \hline \Box \end{array}$$

Draw an array for each. Then write a multiplication sentence.

5. 3 rows of 2

6. 2 rows of 3

_____ × _____ = _____   _____ × _____ = _____

7. The arrays in Exercises 5 and 6 show the _____ Property.

Algebra Write a multiplication sentence with a symbol for the unknown. Then solve.

8. How many ears are on 4 dogs?

    _____ × _____ = ■

    There are _____ ears.

9. There are a total of 16 legs on 2 spiders. How many legs each?

    _____ × ■ = _____

    Each spider has _____ legs.

Write a multiplication sentence.

10.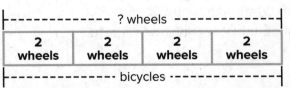

? wheels
2 wheels	2 wheels	2 wheels	2 wheels
bicycles

11.

? buttons
5 buttons	5 buttons
coats

Lesson 2 Multiply by 2 **303**

Problem Solving

Processes &Practices 2 **Use Algebra** Write a multiplication sentence with a symbol for the unknown. Then solve.

12. How many sides are there all together on two squares?

    _____ × 2 = ■

    There are _____ sides.

13. How many gloves are there all together if Ryder has 6 pairs of gloves?

    _____ × 2 = ■

    Ryder has _____ gloves all together.

Brain Builders

14. Paige has 24 magazines in her magazine collection. Each month she adds 2 more magazines to her collection. How many magazines will she have in 3 months? Write two number sentences to show how to solve.

    _____

    She will have _____ magazines.

15. **Processes &Practices 4** **Model Math** Write a word problem about a real-world situation in which a number is multiplied by 2.

    _____
    _____

16. **Building on the Essential Question** What do you notice about all of the products of 2? Use the multiplication table. Identify the pattern.

    _____
    _____

304 Chapter 6 Multiplication and Division Patterns

Name _____

Lesson 2

Multiply by 2

Homework Helper

Need help? connectED.mcgraw-hill.com

Helen buys 2 bunches of bananas. There are 10 bananas in each bunch. How many bananas does Helen buy in all?

Find 2 × 10.

This can be written vertically, also.

Use an array to model 2 groups of 10.

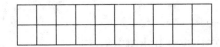

You can write an addition sentence to represent the models.

10 + 10 = 20

OR

You can write a multiplication sentence to represent the models.

2 × 10 = 20

So, Helen bought 20 bananas in all.

Practice

Write an addition sentence and a multiplication sentence.

1.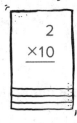

3 groups of 2 is _____.

2 + _____ + _____ = _____

_____ × 2 = _____

2.

4 groups of 2 is _____.

2 + _____ + _____ + _____ = _____

_____ × 2 = _____

Lesson 2 My Homework 305

Draw an array for each. Then write a multiplication sentence.

3. 7 rows of 2

4. 2 rows of 5

____ × ____ = ____   ____ × ____ = ____

Brain Builders

Processes &Practices **2** **Use Algebra** Write a multiplication sentence with a symbol for the unknown. Then solve.

5. Franklin's father gave him and his sister $8 each to buy a ticket for a movie and $2 each for a snack. How much money did Franklin's father give the children all together?

6. There are 5 boys and 2 girls in the Watson family. They all keep their gloves in 1 box in the closet. If each person has a pair of gloves, how many gloves are in the box?

Vocabulary Check

7. Write or draw the meaning of a bar diagram.

8. **Test Practice** James is jumping on a pogo stick. He is counting by twos. If he counts to 12, how many jumps has he made?

Ⓐ 2 jumps   Ⓒ 6 jumps

Ⓑ 4 jumps   Ⓓ 10 jumps

306 **Need more practice?** Download Extra Practice at connectED.mcgraw-hill.com

Name ....

Lesson 3
Divide by 2

ESSENTIAL QUESTION
What is the importance of patterns in learning multiplication and division?

You learned about the division symbol ÷.

Another symbol for division is $\overline{)}$.

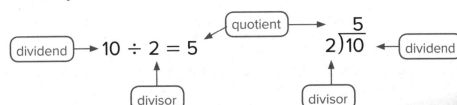

Math in My World

Example 1

Javier and Alexis share an apple equally. If there are 8 slices, how many slices will each of them get?

To share equally between _____ people means to divide by 2. So, find 8 ÷ 2 or $2\overline{)8}$. ← Read: eight divided by two

Partition one counter at a time into each group until the counters are gone. Draw the equal groups at the right.

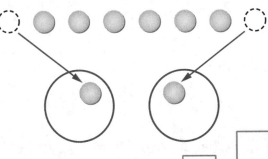

The model shows 8 ÷ 2 = ☐ or $2\overline{)8}$. Each person will get

_____ apple slices.

Online Content at connectED.mcgraw-hill.com

Lesson 3 307

A related multiplication fact can help you find an unknown in a division sentence.

Example 2

Max divided his collection of 12 feathers into 2 groups. How many feathers are in each group? Find the unknown.

Find $12 \div 2 = \blacksquare$ or $2\overline{)12}^{\blacksquare}$

$12 \div 2 = \blacksquare \longrightarrow 2 \times \blacksquare = 12$

A division sentence can be thought of as a multiplication sentence in which you are looking for an unknown factor.

You know that $2 \times 6 = 12$.

So, $12 \div 2 = \boxed{}$ or $2\overline{)12}^{\boxed{}}$. The unknown is $\boxed{}$.

There are _____ feathers in each group.

Guided Practice

Divide. Write a related multiplication fact.

1.

_____ × _____ = _____

2.

$10 \div 2 =$ _____

_____ × _____ = _____

3.

$6 \div 2 =$ _____

_____ × _____ = _____

What are two different ways to find $16 \div 2$?

Name

Independent Practice

Divide. Write a related multiplication fact.

4.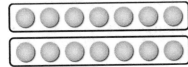

   $14 \div 2 =$ _____

   _____ × _____ = _____

5.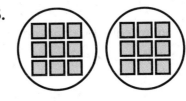

   $2\overline{)18}$

   _____ × _____ = _____

6. $4 \div 2 =$ _____

7. $16 \div 2 =$ _____

8. $18 \div 2 =$ _____

9. $2\overline{)2}$

10. $2\overline{)20}$

11. $2\overline{)6}$

Match the division sentence to the related multiplication sentence.

12. $16 \div 8 = 2$ • $6 \times 2 = 12$

13. $12 \div 2 = 6$ • $2 \times 5 = 10$

14. $10 \div 5 = 2$ • $4 \times 2 = 8$

15. $8 \div 2 = 4$ • $2 \times 8 = 16$

Algebra Find the unknown. Then write a related multiplication sentence.

16. $12 \div 6 = \blacksquare$

    The unknown is _____.

17. $14 \div \blacksquare = 2$

    The unknown is _____.

18. $\blacksquare \div 2 = 3$

    The unknown is _____.

Lesson 3 Divide by 2

Problem Solving

Algebra Write a division sentence with a symbol for the unknown for Exercises 19–20. Then solve.

19. Damian will plant 12 seeds in groups of 2. How many groups of 2 will he have?

20. Kyle and Alan equally divide a package of 14 erasers. How many erasers will each person get?

21. Lydia shared her 16 bottle caps equally with Pilar. Pilar then shared her caps equally with Timothy. How many caps do Pilar and Timothy each have?

Brain Builders

22. **Processes & Practices 6** **Be Precise** You have learned that when any number is multiplied by 2 the product is even. Is the same true for division of an even number divided by 2? Explain. Provide examples to support your reasoning.

23. **Processes & Practices 3** **Find the Error** Blake says that $8 \div 2 = 16$ because $2 \times 8 = 16$. Is Blake correct? Explain.

24. **Building on the Essential Question** How does the relationship between division and multiplication help you find the unknown? How is this relationship like the relationship between addition and subtraction?

310 Chapter 6 Multiplication and Division Patterns

Name

Lesson 3
Divide by 2

Homework Helper   Need help? connectED.mcgraw-hill.com

The school van can carry 12 passengers. There are 2 passengers to a seat. How many seats are in the van?

Find $12 \div 2$, or $2\overline{)12}$.

Partition 12 counters between 2 groups until there are none left.

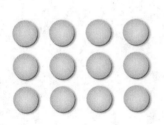  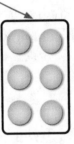

There are 6 counters in each group.

So, $12 \div 2 = 6$ or $2\overline{)12}^{\,6}$. There are 6 seats in the van.

Practice

Divide. Write a related multiplication fact.

1.    2.

$8 \div 2 =$ _____ $18 \div 2 =$ _____

_____ × _____ = _____   _____ × _____ = _____

Divide. Write a related multiplication fact.

3. $20 \div 2 = $ _____

4. $6 \div 2 = $ _____

5. $12 \div 2 = $ _____

6. $2\overline{)8}$

7. $2\overline{)14}$

8. $2\overline{)4}$

Brain Builders

9. **Algebra** In two days, Britt spent a total of $4 at the bead store and $8 at the flower shop. If she spent the same amount each day, how much did she spend? Write a number sentence with a symbol for the unknown. Then solve.

10. **Processes &Practices** **Keep Trying** Ian has 16 red cars and 12 black cars on the floor in his room. If he picks up all the cars and puts an equal number of cars into 2 boxes, how many cars will he put in each box?

Vocabulary Check

11. Write or draw a definition of the word *partition*.

12. **Test Practice** Casey bought a box of 18 granola bars. She ate 2, then shared the rest with her brother. If Casey and her brother have the same number of granola bars, how many of the granola bars did Casey give to her brother?

   Ⓐ 1 granola bar   Ⓒ 9 granola bars
   Ⓑ 8 granola bars   Ⓓ 7 granola bars

Name ....

Lesson 4
Multiply by 5

ESSENTIAL QUESTION
What is the importance of patterns in learning multiplication and division?

You can use patterns to multiply by 5. Multiplying by a number is the same as skip counting by that number.

Math in My World

Example 1

Leandro has 7 nickels. How much money does he have?

One nickel equals 5¢. Skip count by fives to find 7 × 5¢.

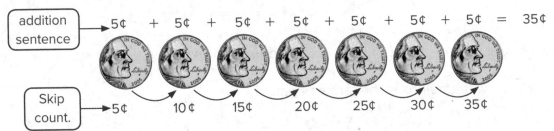

addition sentence → 5¢ + 5¢ + 5¢ + 5¢ + 5¢ + 5¢ + 5¢ = 35¢

Skip count. → 5¢   10¢   15¢   20¢   25¢   30¢   35¢

7 nickels is _____ ¢. 7 × 5¢ = _____ ¢

So, Leandro has _____ ¢.

Notice the pattern in the products.

0 × 5 = **0**
1 × 5 = **5**   ← All of the products end in 0 or 5.
2 × 5 = **10**
3 × 5 = **15**

Helpful Hint
When you multiply by 5, the product will always end in 0 or 5.

Extend the pattern.

4 × 5 = _____

5 × _____ = _____

6 × _____ = _____

7 × _____ = _____

Online Content at connectED.mcgraw-hill.com

Example 2

A watermelon patch has 6 rows of watermelons. Each row has 5 watermelons. How many watermelons are in the farmer's patch? Write a multiplication sentence with a symbol for the unknown.

6 × 5 = ■ ← unknown

 Draw an array with 6 rows.

 Use the Commutative Property to draw another array with 5 rows.

There are _____ rows of _____.

So, 6 × 5 = _____.

The unknown is _____.

There are _____ watermelons in the farmer's patch.

There are _____ rows of _____.

So, 5 × 6 = _____.

The unknown is _____.

Guided Practice

Skip count by fives to find each product. Draw lines to match.

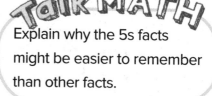
Explain why the 5s facts might be easier to remember than other facts.

1. 4 × 5 = ☐   • 5 + 5 + 5 + 5 + 5 + 5 + 5 + 5

2. 3 × 5 = ☐   • 5 + 5 + 5 + 5

3. 8 × 5 = ☐   • 5 + 5 + 5 + 5 + 5 + 5 + 5

4. 7 × 5 = ☐   • 5 + 5 + 5

314 Chapter 6 Multiplication and Division Patterns

Name _____

Independent Practice

Write an addition sentence to help find each product.

5. 2 × 5 = ____

5 + ____ = ____

6. 3 × 5 = ____

5 + ____ + ____ = ____

7. 7 × 5 = ____

_____ = ____

8. 8 × 5 = ____

_____ = ____

9. 5 × 5 = ____

_____ = ____

10. 9 × 5 = ____

_____ = ____

Draw an array for each. Then write a multiplication sentence.

11. 7 rows of 5

7 × ____ = ____

12. 3 rows of 5

____ × ____ = ____

13. 4 rows of 5

____ × ____ = ____

Algebra Find each unknown. Use the Commutative Property.

14. ■ × 6 = 30
6 × ■ = 30

The unknown is ____.

15. 5 × ■ = 10
■ × 5 = 10

The unknown is ____.

16. 9 × 5 = ■
5 × 9 = ■

The unknown is ____.

Lesson 4 Multiply by 5 **315**

Problem Solving

17. Kai, Lakita, and Maxwell collected acorns. If each gets 5 acorns, how many acorns did they collect? Explain.

18. Processes &Practices 6 **Explain to a Friend** A sunflower costs $6. Evelyn wants to buy 2. Does she have enough money if she has three $5 bills? Explain.

Brain Builders

19. There are 82 members in a band. Part of the band divides into 9 equal groups of 5. How many equal groups of 7 could be made with members that are left?

20. Processes &Practices 2 **Reason** Circle the strategy that will not help you find 6 × 5. Explain.

skip counting   rounding

make an array   draw a picture

21. **Building on the Essential Question** What do you notice about all the products of 5? Use the multiplication table if needed. Which of the products of 5 are also products of 2?

316   Chapter 6 Multiplication and Division Patterns

Name _____

MY Homework

Lesson 4
Multiply by 5

Homework Helper

Need help? connectED.mcgraw-hill.com

There are 6 students. Each student donates $5 to a school fundraiser. How much money did the students donate in all?

Find 6 × $5.

One Way Skip count by fives.

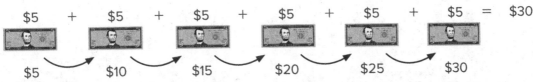

Another Way Draw an array.

6 rows of 5 = 30

So, the 6 students donated a total of $30.

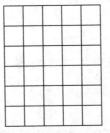

Practice

Write an addition sentence to help find each product.

1. 3 × 5 = _____

_____

2. 8 × 5 = _____

_____

3. 5 × 5 = _____

Lesson 4 My Homework 317

Write a multiplication sentence for each array.

4. 1 row of 5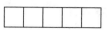

5. 5 rows of 4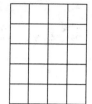

6. 5 rows of 9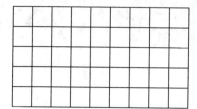

Problem Solving

7. Each pair of tennis shoes costs $25. If Andrea has four $5-bills, does she have enough money to buy one pair? Write a number sentence. Then solve.

Brain Builders

8. For each balloon game you win at the fair, you get 5 tickets. Jamal won 9 balloon games. Gary won 6 balloon games. Do they have enough tickets all together for a prize that is worth 100 tickets? Explain.

9. Processes &Practices 1 **Make Sense of Problems** For a craft, each student will need 5 rubber bands. There are 8 students. Rubber bands come in bags of 9. How many bags will be needed? How many rubber bands will be left over?

10. Test Practice Shawn has 4 nickels. How many walnuts can he buy if he spends all 4 nickels?

Ⓐ 1 walnut   Ⓒ 5 walnuts

Ⓑ 4 walnuts   Ⓓ 20 walnuts

Nuts for Sale
- Peanuts -- 3¢ each
- Walnuts -- 5¢ each
- Chestnuts -- 10¢ each

Lesson 5
Divide by 5

ESSENTIAL QUESTION
What is the importance of patterns in learning multiplication and division?

Use what you know about patterns and multiplying by 5 to divide by 5.

Math in My World

Example 1

A group of 5 friends sold a total of 20 glasses of lemonade. They each sold the same number of glasses. How many glasses of lemonade did they each sell?

Find 20 ÷ 5.

One Way Use counters and partition.

Partition 20 counters into 5 equal groups. Draw the equal groups.

There are _____ counters in each group.

20 ÷ 5 = _____

So, they each sold _____ glasses of lemonade.

Another Way Use repeated subtraction.

Subtract groups of 5 until you reach 0.

Count the number of groups you subtracted.

```
   ①      ②      ③      ④
   20     15     10      5
  − 5    − 5    − 5    − 5
   15     10      5      0
```

Groups of _____ were subtracted _____ times.

There are _____ groups. So, 20 ÷ 5 = _____.

Online Content at connectED.mcgraw-hill.com

Think of division as an unknown factor problem. Use a related multiplication fact.

Example 2

Helpful Hint
Nickels can be used to represent the number 5.

The school store is selling pencils for 5¢ each. If Corey has 45¢, how many pencils can he buy with his money?

Find the unknown in 45¢ ÷ 5¢ = ■ or 5¢)‾45¢

Draw an array. Then use the inverse operation to find the unknown.

Think ■ × 5 = 45   (unknown factor)

You know that _____ × 5 = 45.

So, 45¢ ÷ 5¢ = ☐ or 5¢)‾45¢

The unknown is _____. Corey can buy _____ pencils.

Guided Practice

Use counters to find the number of equal groups or how many are in each group.

Talk MATH
How can you tell if a number is divisible by 5?

1. 35 counters
   5 equal groups
   _____ in each group
   35 ÷ 5 = _____

2. 10 counters
   5 equal groups
   _____ in each group
   10 ÷ 5 = _____

3. Use repeated subtraction to find 30 ÷ 5.

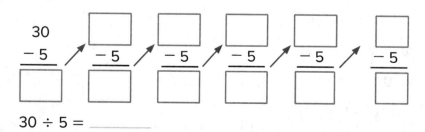

30 ÷ 5 = _____

320 **Chapter 6** Multiplication and Division Patterns

Name _____

Independent Practice

Use counters to find the number of equal groups or how many are in each group.

4. 15 counters
   5 equal groups
   _____ in each group
   $15 \div 5 =$ _____

5. 10 counters
   _____ equal groups
   5 in each group
   $10 \div$ _____ $= 5$

6. 25 counters
   5 equal groups
   _____ in each group
   $25 \div 5 =$ _____

Use repeated subtraction to divide.

7. $10 \div 5 =$ _____

8. $5 \div 1 =$ _____

Algebra Draw an array and use the inverse operation to find each unknown.

9. ■ × 5 = 20
   ? ÷ 4 = 5
   ■ = _____
   ? = _____

10. 5 × ■ = 40
    40 ÷ ? = 8
    ■ = _____
    ? = _____

Use the recipe for Buttermilk Corn Bread. Find how much of each ingredient is needed to make 1 loaf of corn bread.

11. cornmeal _____

12. flour _____

13. eggs _____

14. vanilla extract _____

Buttermilk Corn Bread

10 cups cornmeal
5 cups flour
1 cup sugar
5 Tbsp baking powder
4 tsp salt

3 cups butter
8 cups buttermilk
5 tsp vanilla extract
15 eggs
2 tsp baking soda

Makes: 5 loaves

Lesson 5 Divide by 5   321

Problem Solving

Processes & Practices 2 Use Algebra Write a division sentence with a symbol for the unknown. Then solve.

15. Rose had a 30-inch piece of ribbon. She divided the ribbon into 5 equal pieces. How many inches long is each piece?

16. Garrison collected 45 flags. He displays them in his room in 5 equal rows. How many flags does Garrison have in each row?

Brain Builders

17. **Processes & Practices 1** Keep Trying Addison got 40 points on yesterday's 10-question math quiz. Each question is worth 5 points and there is no partial credit. How many questions did she miss?

18. **Processes & Practices 2** Stop and Reflect Circle the division sentence that does not belong. Explain your reasoning.

$$20 \div 2 = 10 \quad\quad 30 \div 5 = 6$$

$$30 \div 6 = 5 \quad\quad 35 \div 5 = 7$$

19. Building on the Essential Question How can an array help you solve a related multiplication and division problem? Give an example.

Name _____

MY Homework

Lesson 5
Divide by 5

Homework Helper

Need help? connectED.mcgraw-hill.com

Rudy spent $30 for 5 car models. Each model costs the same amount. How much did each car model cost?

Find $30 ÷ 5, or 5)$30.

One Way Use counters and partition.

Partition 30 counters equally among 5 groups until there are none left.

There are 5 equal groups of 6.

Another Way Use repeated subtraction.

Subtract 5 until you get to 0. Count the number of times you subtracted.

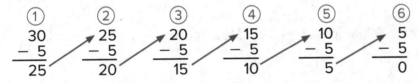

Groups of 5 were subtracted 6 times.

Since $30 ÷ 5 = $6, each model cost $6.

Practice

Partition to find the number of equal groups or how many are in each group.

1. 45 counters

   5 equal groups

   _____ in each group

2. 5 counters

   _____ equal groups

   1 in each group

3. 20 counters

   5 equal groups

   _____ in each group

4. 50 counters

   _____ equal groups

   5 in each group

5. **Algebra** Draw an array and use the inverse operation to find the unknown.

   ■ × 5 = 15

   ? ÷ 3 = 5

   ■ = _____

   ? = _____

Problem Solving

Write a division sentence with a symbol for the unknown for Exercises 6 and 7. Then solve.

6. Antonio scored 40 points on his math test. There were 5 questions on the test, and each was worth the same number of points. How many points did Antonio score for each question?

   _____

Brain Builders

7. Breakfast cost $2. Lunch costs $3. Marcus has $35. How many days can he buy breakfast and lunch?

   _____

8. **Processes &Practices 4** **Model Math** Today 25 girls and 20 boys rode their bikes to school. Each bike rack at school holds 5 bikes. How many bike racks were filled?

   _____

9. **Test Practice** Which number sentence represents this repeated subtraction exercise?

   20 → 15 → 10 → 5
   − 5   − 5   − 5   − 5
   ───   ───   ───   ───
   15   10   5   0

   Ⓐ 20 ÷ 5 = 4   Ⓒ 20 ÷ 2 = 10
   Ⓑ 20 − 20 = 0   Ⓓ 20 − 10 = 10

Check My Progress

Vocabulary Check

Label each with the correct word(s).

bar diagram   factors   partition   product

1.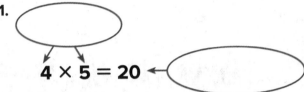

2.

3.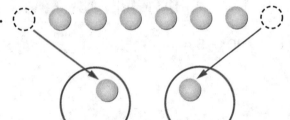

_____ one counter at a time into each group until the counters are gone.

Concept Check

4. Shade the product of the two circled factors. Complete the number sentence.

   6 × 4 = _____

5. Draw a triangle around the product that has the same factors. Write the number sentence that shows the Commutative Property of Multiplication.

   _____

×	0	1	2	3	4	5	6	7	8	9	10
0	0	0	0	0	0	0	0	0	0	0	0
1	0	1	2	3	4	5	6	7	8	9	10
2	0	2	4	6	8	10	12	14	16	18	20
3	0	3	6	9	12	15	18	21	24	27	30
4	0	4	8	12	16	20	24	28	32	36	40
5	0	5	10	15	20	25	30	35	40	45	50
6	0	6	12	18	24	30	36	42	48	54	60
7	0	7	14	21	28	35	42	49	56	63	70
8	0	8	16	24	32	40	48	56	64	72	80
9	0	9	18	27	36	45	54	63	72	81	90
10	0	10	20	30	40	50	60	70	80	90	100

Write an addition sentence and a multiplication sentence for each.

6.  5 groups of 2 is ____

___ + ___ + ___ + ___ + ___ = ___

___ × ___ = ___

7.
	--------- ? pencils ---------	
2 pencils	2 pencils	2 pencils

3 groups of 2 is ____

___ + ___ + ___ = ___

___ × ___ = ___

Divide. Write a related multiplication fact.

8.

6 ÷ 3 = ____

9.

10 ÷ 5 = ____

Brain Builders

10. A postal worker makes 5 trips in the morning and 3 trips in the afternoon to deliver some packages. She carries 2 packages at a time. How many packages are delivered?

11. **Test Practice** Five times as many students bought lunch than packed lunch. If 2 girls and 1 boy packed lunch, which of the following could be used to find how many students bought lunch?

    Ⓐ 5 − 3   Ⓑ 5 × 3   Ⓒ 5 + 3   Ⓓ 5 ÷ 3

326 **Chapter 6** Multiplication and Division Patterns

Name ....

Lesson 6
Problem-Solving Investigation
STRATEGY: Look for a Pattern

ESSENTIAL QUESTION
What is the importance of patterns in learning multiplication and division?

Learn the Strategy

In the first row of her tile pattern, Christina uses 2 tiles. She uses 4 tiles in the second row, 8 tiles in the third row, and 16 tiles in the fourth row. If she continues the pattern, how many tiles will be in the sixth row?

1 Understand

What facts do you know?

There will be _____ tiles in the first row, _____ in the second row, _____ in the third row, and _____ tiles in the fourth row.

What do you need to find?

The number of tiles that will be in row _____.

2 Plan

I will make a table for the information. Then I will look for a pattern.

3 Solve

1st	2nd	3rd	4th	5th	6th
2	4	8	16		

+2   +4   +8   +16   +32

Put the information in a table. Look for a pattern. The numbers double. Now I can continue the pattern. There will be _____ tiles in the sixth row.

4 Check

Does your answer make sense? Explain.

Online Content at connectED.mcgraw-hill.com

Lesson 6   327

Practice the Strategy

Jacy mows lawns every other day. He earns $5 the first day. After that, he earns $1 more than the day before. If he starts mowing on the first day of the month, how much money will he earn the 9th day of the month?

 Understand

What facts do you know?

_____

_____

What do you need to find?

_____

 Plan

_____

_____

3 Solve

 **Check**

Does your answer make sense? Explain.

_____

328 Chapter 6 Multiplication and Division Patterns

Name

Apply the Strategy

Solve each problem by looking for a pattern.

1. A collection of bears is shown. If there are 3 more rows, how many bears are there in all? Identify the pattern.

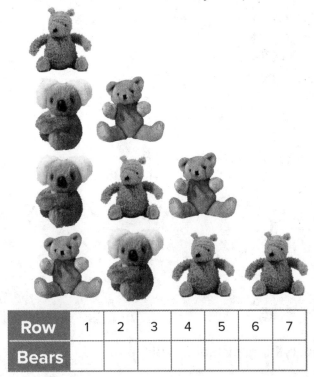

Row	1	2	3	4	5	6	7
Bears							

Brain Builders

2. **Processes &Practices** 8 **Look for a Pattern** Yutaka is planting 15 flowers. He uses a pattern of 1 daisy and then 2 tulips. If the pattern continues, how many tulips will he use? Explain.

Daisies	Tulips	Total
1	2	
2	4	
3		

Lesson 6 Problem-Solving Investigation 329

Review the Strategies

For Exercises 3–5, use the sign shown.

HEALTH SHACK'S SNACKS

Sunflower seeds ...... 10¢ per package
Dried fruit .... 10 pieces for 50¢
Juice .... 20¢ each
Yogurt .... 2 for 80¢

Use any strategy to solve each problem.
- Use an estimate or exact answer.
- Make a table.
- Look for a pattern.
- Use models.

3. Julius spent 70¢ on sunflower seeds. How many packages did he buy?

_____

4. How much did Neila pay for 1 yogurt?

_____

5. How much would it cost to buy 1 of everything, including 1 piece of dried fruit?

_____

6. Orlon collected 40 comic books. He keeps 10 comic books for himself and divides the rest equally among his 5 friends. How many comic books does each friend get?

_____

7. **Processes & Practices 5** **Use Math Tools** The amount of light given off by a light bulb is measured in lumens. Each of the 2 light bulbs in Hudson's room gives 1,585 lumens of light. Together, about how many lumens do both light bulbs give off?

_____

Name

Lesson 6

Problem Solving:
Look for a Pattern

Homework Helper Need help? connectED.mcgraw-hill.com

At the arcade, Kelly began with 48 tokens. She gave 24 to Kuri. Then she gave 12 to Tonya. If this pattern continues, how many tokens will Kelly give away next?

1 Understand

What facts do you know?
Kelly began with 48 tokens.
She gave away 24 tokens, then 12 tokens.

What do you need to find?
how many tokens Kelly will give away next

2 Plan

I will look for a pattern.

3 Solve

The pattern is 48, 24, 12, . . .

Each number is half as much as the one before it.

The pattern is divide by 2.

$12 \div 2 = 6$

So, Kelly will give away 6 tokens next.

4 Check

Does the answer make sense?
Half of 12 is 6. The answer makes sense.

Lesson 6 My Homework 331

Problem Solving

Solve each problem by looking for a pattern.

1. Adam is lining up his toy train cars. If he continues this color pattern, what color will the 18th car be?

2. Marissa delivers newspapers on Highview Drive. The first house number is 950, the next is 940, and the third is 930. If the pattern continues, what will the next house number be?

Brain Builders

3. Kyle is training for a bike race. He rides 5 miles one day, 10 miles the next day, and 15 miles the third day. If Kyle continues this pattern, what is the total distance he will have ridden after 5 days?

4. The Hornets basketball team won their first game by 18 points, their second game by 15 points, and their third game by 12 points. If the pattern continues, by how many points will they win their fifth game?

5. Darcy wears brown pants to work one day, blue pants the next day, and a skirt on the third day. If this pattern continues every three days, what will she wear to work on the seventh day? Explain how you know.

Name _____

Lesson 7
Multiply by 10

ESSENTIAL QUESTION
What is the importance of patterns in learning multiplication and division?

Math in My World

Example 1

Orlando found 8 dimes. How much money did Orlando find?
A dime equals 10¢. Count by tens to find 8 × 10¢.

addition sentence → 10¢ + 10¢ + 10¢ + 10¢ + 10¢ + 10¢ + 10¢ + 10¢ = 80¢

Skip count → 10¢ 20¢ 30¢ 40¢ 50¢ 60¢ 70¢ 80¢

8 dimes is _____ ¢. 8 × 10¢ = _____ ¢

So, Orlando found _____ ¢.

Notice the pattern in the products.

1 × 10 = **10** ← The ones digit of the product is zero.
2 × 10 = **20**
       same

Extend the pattern.

3 × 10 = _____

4 × 10 = _____

5 × _____ = _____

6 × _____ = _____

7 × _____ = _____

8 × _____ = _____

Helpful Hint
When you multiply by 10, the product will always end in 0.

Online Content at connectED.mcgraw-hill.com

Lesson 7 333

Example 2

Andrew saw footprints on the beach. He counted 10 toes on each of 3 sets of footprints. How many toes did Andrew count in all? Write a multiplication sentence with a symbol for the unknown.

3 × 10 = ■ ← unknown

Skip count on a number line. Count three equal jumps of 10.

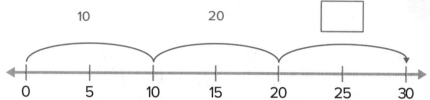

The number line shows that 3 × 10 = _____. The unknown is _____.

So, Andrew counted _____ toes in the sand.

Guided Practice

Skip count by tens to find each product. Draw lines to match.

1. 5 × 10 = ☐   • 10 + 10

2. 2 × 10 = ☐   • 10

3. 7 × 10 = ☐   • 10 + 10 + 10 + 10 + 10 + 10 + 10

4. 1 × 10 = ☐   • 10 + 10 + 10 + 10 + 10

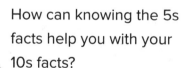

How can knowing the 5s facts help you with your 10s facts?

5. Complete the pattern.

10, 20, _____, _____, 50, 60, 70, _____, _____, 100

334 Chapter 6 Multiplication and Division Patterns

Name _____

Independent Practice

Skip count to find each product. Write the addition sentence.

6. 4 × 10 = ____   10 + ____ + ____ + ____ = ____

7. 6 × 10 = ____   10 + ____ + ____ + ____ + ____ + ____ = ____

8. 3 × 10 = ____   ____ + ____ + ____ = ____

9. 5 × 10 = ____   ____ + ____ + ____ + ____ + ____ = ____

Algebra Use the number line to find each unknown.

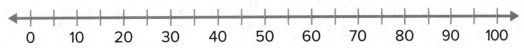

10. ■ × 6 = 60
    6 × ■ = 60

11. 10 × ■ = 10
    ■ × 10 = 10

12. 9 × 10 = ■
    10 × 9 = ■

The unknown is ____.   The unknown is ____.   The unknown is ____.

Multiply.

13. 10 × 2 = ____   14. 10 × 6 = ____   15. 10 × 5 = ____

16. 10
  × 3

17. 9
  × 10

18. 10
  × 1

Use the Commutative Property to find each product. Draw a line to match.

19. 8 × 10 = ☐

20. 10 × 5 = ☐

21. 6 × 10 = ☐

• 10 × 6 = 60

• 10 × 8 = 80

• 5 × 10 = 50

Lesson 7 Multiply by 10 335

Problem Solving

Some of the world's largest glass sculptures are found in the United States. Use the clues in Exercises 22–25 to find the length of each sculpture.

World's Largest Glass Sculptures	
Sculpture Name	Length (feet)
Fiori di Como, NV	?
Chihuly Tower, OK	?
Cobalt Blue Chandelier, WA	?
River Blue, CT	?

22. Fiori di Como: 5 less than 7 × 10

_____

23. Chihuly Tower: 5 more than 10 × 5

_____

24. Cobalt Blue Chandelier: 9 more than 2 × 10

_____

25. River Blue: 4 more than 10 × 1

_____

Brain Builders

26. **Processes &Practices 2** **Use Number Sense** There are 5 giraffes and 10 birds. How many legs are there all together? Use number models to support your answer.

_____

27. **Processes &Practices 2** **Reason** Explain how you know that a multiplication sentence with a product of 25 cannot be a 10s fact. What kind of fact is it? Explain.

_____

28. **Building on the Essential Question** How can I use patterns to multiply numbers by 10? What is the relationship between 10 and the product?

_____

Name ....

MY Homework

Lesson 7

Multiply by 10

Homework Helper

Need help? connectED.mcgraw-hill.com

There are 8 players on the tennis team. Each family contributes $10 toward a gift for the coach. What is the total amount collected for the coach's gift?

Find 8 × $10.

Skip count by tens.

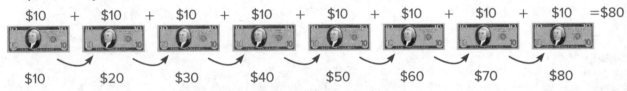

So, the total amount collected from 8 families was $80.

Practice

Skip count by tens to find each product. Write the addition sentence.

1. 5 × 10 = _____

_____

2. 2 × 10 = _____

_____

3. 7 × 10 = _____

_____

4. 3 × 10 = _____

_____

Lesson 7 My Homework 337

Algebra Use the number line to find each unknown.

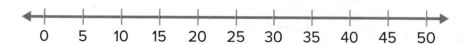

5. ■ × 4 = 40

   4 × ■ = 40

   The unknown is _____.

6. 10 × ■ = 20

   ■ × 10 = 20

   The unknown is _____.

7. 10 × ■ = 50

   ■ × 10 = 50

   The unknown is _____.

Problem Solving

For Exercises 8–10, write number models to support your answer.

8. Fiona's class went on a field trip to the art museum. The class rode in vans with 10 people in each van. How many people went on the field trip if they took 4 full vans?

_____

Brain Builders

9. **Processes &Practices 5 Use Math Tools** During the first half of the football game, Carlos ran with the ball 3 times. Each time, he ran 10 yards. During the second half, he ran 2 times. Each time, he ran 6 yards. How many yards did Carlos run all together?

_____

10. Each time Allison goes to the recycling center, she takes 10 bags of cans. She will go twice this month, 3 times next month, and once the following month. How many bags of cans will Allison take to the recycling center in these three months?

_____

11. **Test Practice** Byron has 70 pennies. He stacks them in groups of 10. How many stacks of pennies can Byron make?

    Ⓐ 7 stacks   Ⓑ 9 stacks   Ⓒ 8 stacks   Ⓓ 10 stacks

Name _____

Lesson 8
Multiples of 10

ESSENTIAL QUESTION
What is the importance of patterns in learning multiplication and division?

The product of a given number, such as 10, and any other number is a **multiple**. You can use a basic fact and patterns of zeros to mentally find multiples of 10.

Math in My World

Be our guest!

Example 1

A new hotel has 3 floors. There are 20 rooms on each floor. What is the total number of rooms in the hotel?

Find 3 × 20. ← 20 is a multiple of 10, since 2 × 10 = 20

One Way Use a basic fact and patterns.

3 × 2 = 6 ← basic fact

3 × 20 = _____ ← 3 × 2 = 6, so 3 × 20 = 60

Another Way Use place value.

Think of 3 × 20 as 3 × 2 tens.

Use base-ten blocks to model 3 equal groups of 2 tens. Draw your model at the right.

3 × 2 tens = _____ tens

So, 3 × 20 = _____ .

6 tens = 60

Check for Reasonableness
Use repeated addition.

20 + _____ + _____ = _____

Online Content at connectED.mcgraw-hill.com

Properties can be used to multiply a number by a multiple of 10.

Example 2

Ellie is buying 2 bags of beads to add to her bead collection. Each bag has 40 beads. How many beads is Ellie buying?

Find 2 × 40.

2 × 40 = 2 × (4 × 10)   Write 40 as 4 × 10.

= (2 × 4) × 10   Find 2 × 4 first.

= _____ × 10   Multiply.

= _____

So, Ellie is buying _____ beads.

Helpful Hint
The way in which numbers are grouped does not change their product.

Example 3

Find the unknown in 4 × 50 = ■.

4 × 50 = 200   4 × 5 tens = 20 tens

Sometimes the basic fact has a zero. Keep that zero, then add the other zero.

So, 4 × 50 = _____. The unknown is _____.

Talk MATH

Find the product of 3 × 20 and 2 × 30. What do you notice about the products? Is this an example of the Commutative Property of Multiplication? Explain.

Guided Practice

Multiply. Use place value.

1. 2 × 20 = 2 × _____ tens

    = _____ tens

    So, 2 × 20 = _____.

2. 5 × 60 = 5 × _____ tens

    = _____ tens

    So, 5 × 60 = _____.

340 Chapter 6 Multiplication and Division Patterns

Name _____

Independent Practice

Multiply. Use a basic fact.

3. $5 \times 5 =$ _____

   So, $5 \times 50 =$ _____.

4. $6 \times 2 =$ _____

   So, $6 \times 20 =$ _____.

5. $5 \times 7 =$ _____

   So, $5 \times 70 =$ _____.

Multiply. Use place value.

6. $5 \times 20 =$

   _____ $\times$ _____ tens = _____ tens

   So, $5 \times 20 =$ _____.

7. $2 \times 70 =$

   _____ $\times$ _____ tens = _____ tens

   So, $2 \times 70 =$ _____.

8. $8 \times 50 =$

   _____ $\times$ _____ tens = _____ tens

   So, $8 \times 50 =$ _____.

9. $2 \times 80 =$

   _____ $\times$ _____ tens = _____ tens

   So, $2 \times 80 =$ _____.

Multiply to find each product. Draw lines to match.

10. $2 \times 90 =$ _____

11. $5 \times 40 =$ _____

12. $5 \times 90 =$ _____

• $5 \times (4 \times 10) = (5 \times 4) \times 10$
   $= 20 \times 10$
   $=$ _____

• $5 \times (9 \times 10) = (5 \times 9) \times 10$
   $= 45 \times 10$
   $=$ _____

• $2 \times (9 \times 10) = (2 \times 9) \times 10$
   $= 18 \times 10$
   $=$ _____

Algebra Find each unknown.

13. $2 \times \blacksquare = 100$

    The unknown is _____.

14. $2 \times \blacksquare = 60$

    The unknown is _____.

15. $6 \times 50 = \blacksquare$

    The unknown is _____.

Problem Solving

Write a multiplication sentence with a symbol for the unknown for Exercises 16–17. Then solve.

16. **Processes &Practices 2** **Use Algebra** Demont's card album has 20 pages, and 6 trading cards are on each page. How many cards are there in all?

17. There are 90 houses with 10 windows each. How many windows are there in all?

Brain Builders

18. Carlita collected 2 boxes of teddy bears. Each box holds 20 bears. She sells each bear for $2. How much money did she earn?

19. **Processes &Practices 4** **Model Math** Write a multiplication sentence that uses a multiple of 10 and has a product of 120.

20. **Processes &Practices 8** **Look for a Pattern** Describe the pattern you see when multiplying 5 × 30.

What is the product of 5 × 300? Describe the pattern.

21. **Building on the Essential Question** How do basic facts and patterns help me multiply a number by a multiple of 10?

342 Chapter 6 Multiplication and Division Patterns

Name _____

MY Homework

Lesson 8

Multiples of 10

Homework Helper

Need help? connectED.mcgraw-hill.com

There are 3 shelves in the cabinet. Each shelf holds 40 cans. How many cans will fit in the cabinet?

You need to find 3×40.

One Way Use a basic fact and patterns.

$3 \times 4 = 12$ ← basic fact

$3 \times 40 = 120$ ← pattern

Another Way Use place value.

Use base-ten blocks to model 3 groups of 4 tens.

3×4 tens = 12 tens; 12 tens = 120.
So, $3 \times 40 = 120$.

Use repeated addition to check:
$40 + 40 + 40 = 120$

So, 120 cans will fit in the cabinet.

Practice

Multiply. Use place value.

1. $2 \times 40 =$

   $2 \times$ ____ tens = ____ tens

   So, $2 \times 40 =$ ____.

2. $5 \times 60 =$

   $5 \times$ ____ tens = ____ tens

   So, $5 \times 60 =$ ____.

3. $5 \times 30 =$

   $5 \times$ ____ tens = ____ tens

   So, $5 \times 30 =$ ____.

4. $10 \times 20 =$

   $10 \times$ ____ tens = ____ tens

   So, $10 \times 20 =$ ____.

Multiply. Use a basic fact.

5. 10 × 3 = _____

   So, 10 × 30 = _____

6. 2 × 9 = _____

   So, 2 × 90 = _____

7. 2 × 8 = _____

   So, 2 × 80 = _____

8. 5 × 5 = _____

   So, 5 × 50 = _____

Brain Builders

Write a multiplication sentence to solve.

9. Harlan has 3 women's and 2 men's antique watches. Each watch has a value of $90. How much are Harlan's watches worth?

   _____

10. **Processes & Practices 1** **Keep Trying** Trey uses 40 nails to put up the frame around each window. There are 5 windows in the bedroom and 1 window in the bathroom. How many nails will Trey use?

    _____

11. Chloe uses 80 candy wrappers to make a paper necklace. She is making necklaces for herself and 9 friends. How many candy wrappers will Chloe need?

    _____

Vocabulary Check

12. Circle the number sentence that shows 20 is a multiple of 2.

    2 × 10 = 20   2 + 10 = 12

    2 × 5 = 10   10 ÷ 2 = 5

13. **Test Practice** Which is equal to 52 tens?

    Ⓐ 52,010   Ⓑ 5,210   Ⓒ 5,200   Ⓓ 520

Name ....

Lesson 9
Divide by 10

ESSENTIAL QUESTION
What is the importance of patterns in learning multiplication and division?

Math in My World

Example 1

The third grade class needs 50 juice bars. How many boxes of juice bars will they need if there are 10 bars in each box?

Find 50 ÷ 10.

One Way Use a number line.

Start at 50 and count back by 10s.

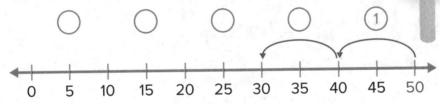

Groups of 10 were counted back _____ times.

Another Way Use repeated subtraction.

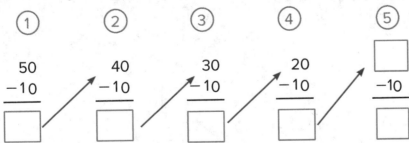

Subtract groups of _____ until you reach _____.

Groups of _____ were subtracted _____ times.

Either way, 50 ÷ 10 = _____.

The third grade class will need _____ boxes of juice bars.

Online Content at connectED.mcgraw-hill.com

Lesson 9 345

Think of division as an unknown factor problem.
Use a related multiplication fact.

Example 2

A quarterback threw the football for a total of 70 yards. Each time he threw the ball, the team gained 10 yards. How many throws did the quarterback make? Write a division sentence with a symbol for the unknown.

Find $70 \div 10 = \blacksquare$.

You know that $10 \times$ _____ $= 70$. The unknown factor is _____.

Since division and multiplication are inverse operations,

$70 \div 10 =$ _____

The unknown is _____.

The quarterback made _____ throws.

Talk MATH
When you divide by 10, what do you notice about the quotient and the dividend?

Guided Practice

Use repeated subtraction to divide.

1. $90 \div 10 =$ _____

2. $40 \div 10 =$ _____

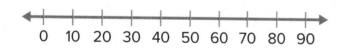

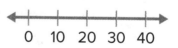

3. $60 \div 10 =$ _____

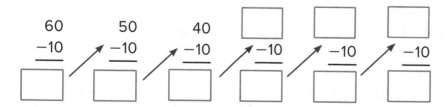

346 Chapter 6 Multiplication and Division Patterns

Name _____

Independent Practice

Use repeated subtraction to divide.

4. 20 ÷ 10 = _____

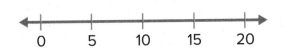

5. 10 ÷ 10 = _____

6. 30 ÷ 10 = _____

```
  30
- 10
─────
```

7. 80 ÷ 10 = _____

```
  80
- 10
─────
```

Algebra Use a related multiplication fact to find each unknown.

8. 50 ÷ 10 = ■

10 × _____ = 50

The unknown is _____ .

9. 70 ÷ ■ = 7

_____ × 7 = 70

The unknown is _____ .

10. 90 ÷ 10 = ■

10 × _____ = 90

The unknown is _____ .

11. 60 ÷ ■ = 6

_____ × 6 = 60

The unknown is _____ .

12. 100 ÷ 10 = ■

10 × _____ = 100

The unknown is _____ .

13. ■ ÷ 10 = 4

10 × 4 = _____

The unknown is _____ .

Lesson 9 Divide by 10 347

Problem Solving

Algebra For Exercises 14–15, write a division sentence with a symbol for the unknown. Then solve.

14. Ken wants to divide 40 flowers equally into 10 vases. How many flowers go in each vase?

_____

15. Rona saw 60 cars at a car show. If he saw 10 of each kind of car, how many different kinds of cars were there?

_____

Brain Builders

16. **Processes &Practices 5** **Use Math Tools** The table shows the amount of money each child has saved in $10-bills.

What is the difference in the least amount of money saved and the greatest amount of money saved?

Savings Accounts	
Name	Saved
Rebecca	$70
Bret	$30
Monsa	$80
Hakeem	$90

_____

How many $10-bills is the difference equal to?

_____

How many $10-bills have the children saved together?

_____

17. **Processes &Practices 2** **Use Number Sense** Use the numerals 0, 7, and 8 to write two 2-digit numbers that can each be divided by 10. The numerals can be used more than one time.

_____

18. **Building on the Essential Question** How can skip counting by 10s help you find the quotient of 10s facts?

_____

348 Chapter 6 Multiplication and Division Patterns

Name _____

Lesson 9
Divide by 10

Homework Helper

Need help? connectED.mcgraw-hill.com

Ms. Mickle's classroom has 30 desks with 10 desks in each row. How many rows of desks are there?

Find 30 ÷ 10.

Subtract groups of 10 until you reach 0.

One Way Use a number line.

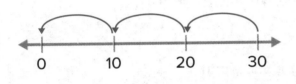

Another Way Use repeated subtraction.

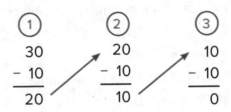

3 groups of 10 were subtracted and you know that 10 × 3 = 30.

So, 30 ÷ 10 = 3. There are 3 rows of desks.

Practice

Use repeated subtraction to divide.

1. 70 ÷ 10 = _____

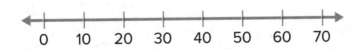

2. 60 ÷ 10 = _____

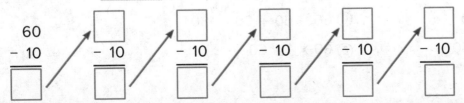

Lesson 9 My Homework 349

Algebra Use a related multiplication fact to find each unknown.

3. $80 \div 10 = \blacksquare$

   $10 \times \underline{} = 80$

   The unknown is $\underline{}$.

4. $\blacksquare \div 10 = 3$

   $10 \times 3 = \underline{}$

   The unknown is $\underline{}$.

5. $\blacksquare \div 10 = 10$

   $10 \times 10 = \underline{}$

   The unknown is $\underline{}$.

6. $20 \div 10 = \blacksquare$

   $10 \times \underline{} = 20$

   The unknown is $\underline{}$.

Brain Builders

7. Morgan has 40 cents in her pocket and 50 cents in her purse. All of the change is in dimes. How many dimes does Morgan have in all?

8. Ricky spent $90 at the supermarket. He bought $30 worth of fruit. He spent the rest of the money on steaks. If he bought 10 steaks and they each cost the same amount, what was the price of each steak?

9. **Processes & Practices** **1** **Make Sense of Problems** Annie bought a bag of 80 mini-carrots. She eats 5 carrots with lunch each day and eats another 5 each night as a snack. In how many days will the bag of carrots be gone?

10. **Test Practice** Bill has a collection of 60 books he wants to donate to the library. Which number sentence shows how Bill can divide the books equally as he packs them in boxes?

    Ⓐ $60 \div 6 = 10$

    Ⓑ $60 - 10 = 50$

    Ⓒ $60 + 60 + 60 = 180$

    Ⓓ $60 \times 1 = 60$

Need more practice? Download Extra Practice at connectED.mcgraw-hill.com

Name _____

Fluency Practice

Processes & Practices 6

Multiply.

1. 2 × 9 = ____
2. 5 × 3 = ____
3. 2 × 4 = ____
4. 10 × 6 = ____

5. 2 × 3 = ____
6. 2 × 5 = ____
7. 2 × 2 = ____
8. 5 × 1 = ____

9. 5 × 4 = ____
10. 2 × 6 = ____
11. 2 × 7 = ____
12. 10 × 2 = ____

13. 10
   × 3
   ___

14. 5
   × 6
   ___

15. 2
   × 8
   ___

16. 10
   × 4
   ___

17. 5
   × 7
   ___

18. 5
   × 5
   ___

19. 10
   × 6
   ___

20. 5
   × 8
   ___

21. 2
   × 1
   ___

22. 5
   × 2
   ___

23. 5
   × 9
   ___

24. 10
   × 5
   ___

Online Content at connectED.mcgraw-hill.com

Name

Fluency Practice

Divide.

1. $10 \div 5 =$ _____
2. $20 \div 5 =$ _____
3. $30 \div 10 =$ _____
4. $8 \div 2 =$ _____

5. $16 \div 2 =$ _____
6. $50 \div 10 =$ _____
7. $35 \div 5 =$ _____
8. $25 \div 5 =$ _____

9. $45 \div 5 =$ _____
10. $60 \div 10 =$ _____
11. $40 \div 5 =$ _____
12. $10 \div 2 =$ _____

13. $2\overline{)12}$
14. $5\overline{)30}$
15. $10\overline{)20}$
16. $5\overline{)15}$

17. $10\overline{)70}$
18. $2\overline{)14}$
19. $2\overline{)18}$
20. $5\overline{)5}$

21. $10\overline{)40}$
22. $2\overline{)20}$
23. $2\overline{)6}$
24. $2\overline{)4}$

352 Chapter 6 Multiplication and Division Patterns

Review

Chapter 6
Multiplication and Division Patterns

Vocabulary Check

Use the clues and word bank below to complete the crossword puzzle.

bar diagram   factor   multiple

partition   product

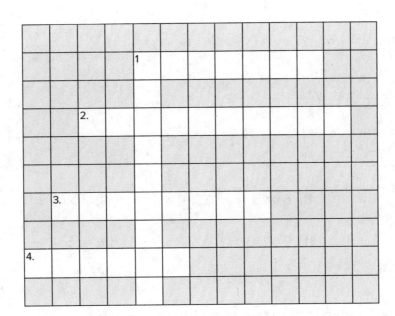

Across

1. The answer to a multiplication problem.
2. A drawing that helps to organize your information.
3. The product of a given number and any other whole number.
4. A number that is multiplied by another number.

Down

1. To separate a number of objects into equal groups.

My Chapter Review 353

Concept Check

Write an addition sentence and a multiplication sentence. Then draw an array.

5. 2 rows of 3 is _____ .

_____ + _____ = _____

_____ × _____ = _____

6. 5 rows of 2 is _____ .

_____ + _____ + _____ + _____ = _____

_____ × _____ = _____

Multiply.

7. 7 × 10 = _____

8. 6 × 5 = _____

9. 1 × 5 = _____

10. 2 × 10 = _____

11. 9 × 10 = _____

12. 4 × 5 = _____

Algebra Use a related multiplication fact to find each unknown.

13. 30 ÷ 10 = ■

_____ × 10 = 30

The unknown is _____ .

14. 60 ÷ ■ = 6

_____ × 6 = 60

The unknown is _____ .

15. 40 ÷ 5 = ■

_____ × 5 = 40

The unknown is _____ .

Find each product or quotient.

16. 5 ÷ 1 = _____

17. 50 ÷ 10 = _____

18. 9 × 5 = _____

19. 30 × 2 = _____

20. 2 × 80 = _____

21. 5 × 70 = _____

Problem Solving

Algebra Write a number sentence with a symbol for the unknown for Exercises 22 and 23. Then solve.

22. Lucia's classroom has tables that are a total of 25 feet long. If there are 5 tables, how long is each table?

_____

23. Julian bought 8 books this year. He gets a free book every time he buys 1. How many books all together did he get this year?

_____

Brain Builders

24. Robert wrote this division sentence.

$$20 \div 2 = 10$$

Write two number sentences that he could use to check his work.

_____

25. The table below shows the cost of movie tickets.

Movie Tickets					
Input	1	2	3	4	5
Output	$8	$16	$24	?	?

What is the cost for 5 movie tickets? 7 movie tickets? 10 movie tickets?

_____

26. **Test Practice** When Xavier buys his lunch, he saves the nickel he gets as change. How much money will Xavier have if he buys his lunch 4 times one week and 2 times the second week?

  Ⓐ 5¢   Ⓒ 25¢

  Ⓑ 6¢   Ⓓ 30¢

Reflect

Chapter 6
Answering the ESSENTIAL QUESTION

Use what you learned about multiplication and division patterns to complete the graphic organizer.

ESSENTIAL QUESTION

What is the importance of patterns in learning multiplication and division?

Multiplication Patterns	Division Patterns

Reflect on the ESSENTIAL QUESTION Write your answer below.

_____

_____

_____

356 Chapter 6 Multiplication and Division Patterns

Name _____ Date _____

Score _____

Performance Task

Brain Builders

Car Wash

David runs a car wash. He charges different amounts based on the size of the vehicle and how much cleaning the customer wants.

Show all your work to receive full credit.

Part A

David charges $10 to vacuum any regular-size car or SUV. Today he vacuumed 8 cars. How much money did David earn for these cleanings? Write a multiplication sentence to support your answer.

Part B

David can do a super-duper wash on any car. He can do 2 super-duper washes each hour. If 14 cars choose the super-duper wash today, for how many hours will David work? Write a division sentence to support your answer.

Online Content at connectED.mcgraw-hill.com

Performance Task 356PT1

Part C

David charges $4 to wash all the windows of a car, inside and out. The amount of money he earns washing the windows must end in one of what digits? Why? List all the possible digits.

Part D

David wears waterproof overalls each day he works at the car wash. He always wears blue overalls on the first day of the month, yellow overalls on the second day, red overalls on the third day, white overalls on the fourth day, green overalls on the fifth day, and brown overalls on the sixth day. If this pattern is repeated, what color of overalls will he wear on the fifteenth day?

Part E

David charges $5 to wash a motorcycle. If David made $45 washing motorcycles on one day, how many motorcycles did he wash? Write a division sentence to support your answer.

Chapter 7 Multiplication and Division

ESSENTIAL QUESTION

What strategies can be used to learn multiplication and division facts?

My Fun Friends

Watch a video!

357

Name _____

MY Chapter Project

Plant an Array

1. Write your group's multiplication sentence.

_____

2. Work together to determine the arrays that can be made using your multiplication sentence. Choose one array and draw it in the space below.

3. Fill the plastic cups with soil. Plant your seeds or beans in the cups. Arrange the cups in the array shown above.

4. When the plants start to sprout, take a "garden tour." Below, list all the multiplication sentences you saw on the tour. Draw an array to go with each multiplication sentence.

358 Chapter 7 Multiplication and Division

Name _____

Am I Ready?

Tell whether the groups in each pair are equal.

1. 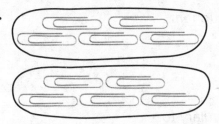 _____

2. 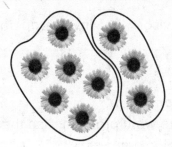 _____

Use the array to complete each pair of number sentences.

3. $2 \times$ _____ $= 8$
   $8 \div$ _____ $= 4$

4. $1 \times 3 =$ _____
   $3 \div$ _____ $= 3$

Draw lines to match the division sentence to the related multiplication sentence.

5. $6 \div 2 = 3$ • $5 \times 3 = 15$
6. $15 \div 3 = 5$ • $8 \times 2 = 16$
7. $20 \div 4 = 5$ • $4 \times 5 = 20$
8. $16 \div 2 = 8$ • $3 \times 2 = 6$

9. **Algebra** Mrs. June wants to divide 30 folders equally among 10 students. How many folders will each student get? Write a division sentence with a symbol for the unknown. Then solve.

Shade the boxes to show the problems you answered correctly.

How Did I Do? 1 2 3 4 5 6 7 8 9

Online Content at connectED.mcgraw-hill.com

Name ....

MY Math Words

Review Vocabulary

dividend   divisor   inverse operations   quotient

Making Connections
Read and solve the word problem. Use the review vocabulary to complete the graphic organizer.

Our class of 20 students will go on a field trip to the zoo. We will need to divide equally into 5 cars. How many students will ride in each car?

20 ÷ 5 = _____

- 20 students _____
- 5 cars _____
- 4 students in each car _____
- 20 ÷ 5 = 4
  4 × 5 = 20
  _____

360 Chapter 7 Multiplication and Division

MY Vocabulary Cards

Lesson 7-3

decompose

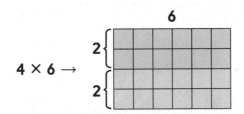

Lesson 7-7

Identity Property of Multiplication

$5 \times 1 = 5$

Lesson 7-3

known fact

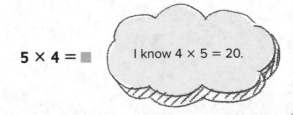

Lesson 7-7

Zero Property of Multiplication

$7 \times 0 = 0$

$70 \times 0 = 0$

$700 \times 0 = 0$

Ideas for Use

- Use the blank cards to write your own vocabulary cards.
- Write this chapter's essential question on the front of a blank card. Write words or phrases from the lessons that help you answer the essential question.

When any number is multiplied by 1, the product is that number.

How can this property help you solve multiplication problems with a factor of 1?

To separate into parts.

Explain how decomposing into known facts can help make a difficult math fact easier to solve.

When any number is multiplied by 0, the product is zero.

How can this property help you solve multiplication problems with a factor of 0?

A fact that you know by memory.

When might you use a known multiplication fact?

MY Foldable

FOLDABLES Follow the steps on the back to make your Foldable.

Divide 24 by 3

Find $3\overline{)24}$

■ × 3 = 24

The unknown is _____.

① 24 − 3 = 21
② 21 − 3 = 18
③ 18 − ___
○ 15 − 3
○ 12

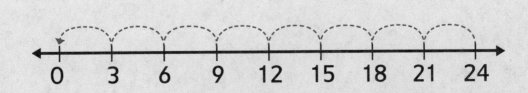

   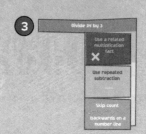

Three Ways to divide 24 by 3

Use a related multiplication fact

✗

Use repeated subtraction

Skip count backwards on a number line

Name _____

Lesson 1
Multiply by 3

ESSENTIAL QUESTION
What strategies can be used to learn multiplication and division facts?

 Math in My World

Example 1

There are 3 dogs. Each dog buried 4 bones in a yard. How many bones are buried in the yard?

Find 3 × 4.

Write it like this, also. $\begin{array}{r} 3 \\ \times 4 \\ \hline \end{array}$

One Way Use an array.

Find 3 rows of 4 bones.
The array shows that 3 × 4 = _____.

There are _____ bones buried in the yard.

Use the Commutative Property to write another multiplication sentence for this array.

_____ × _____ = _____

Another Way Use a number line.

Skip count to find 3 groups of 4.

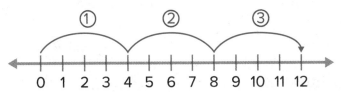

The number line shows _____ jumps of 4 is _____.

So, 3 × 4 = _____. There are _____ bones buried in the yard.

Online Content at connectED.mcgraw-hill.com

Lesson 1 365

Example 2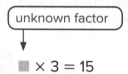

Luke bought 15 packs of seeds. How many different kinds of seeds does he have if there are 3 of each kind? Find the unknown factor. Use a related multiplication fact.

Helpful Hint
How many groups of 3 equal 15?

unknown factor
↓
■ × 3 = 15

You know that 3 × 5 = 15.

So, the _____ Property tells you

that _____ × 3 = 15, also.

The unknown factor is _____.

Check

3 + 3 + 3 + 3 + 3 = 15

Guided Practice

1. Draw an array. Then write two multiplication sentences.

   3 rows of 2

   _____

   _____

2. Circle the number sentence that is represented by the number line.

   15 × 1 = 15   3 × 5 = 15   5 + 10 = 15

Talk MATH

Explain two strategies you can use to find the product of 7 × 3.

366 Chapter 7 Multiplication and Division

Name _____

Independent Practice

Draw an array for each. Then write two multiplication sentences.

3. 3 rows of 4

   _____

   _____

4. 7 rows of 3

   _____

   _____

5. 3 rows of 8

   _____

   _____

6. 5 rows of 3

   _____

   _____

Draw jumps on the number line to find each product.

7.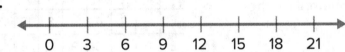

   $3 \times 6 =$ _____

8.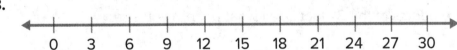

   $3 \times 9 =$ _____

Algebra Find the unknown factor. Use the Commutative Property.

9. ■ $\times 3 = 24$

   $3 \times$ ■ $= 24$

   The unknown is _____.

10. $3 \times$ ■ $= 15$

    ■ $\times 3 = 15$

    The unknown is _____.

11. $3 \times$ ■ $= 6$

    ■ $\times 3 = 6$

    The unknown is _____.

Lesson 1 Multiply by 3

Problem Solving

12. There are 9 singers. They each perform 3 songs at the recital. How many songs were performed?

_____

13. There are 7 daisies and 7 tulips. Each flower has 3 petals. How many petals are there in all?

_____

14. Henry, Jaime, and Kayla each had 3 snacks packed in their lunch boxes. They each ate one snack in the morning. How many snacks are left in all?

_____

Brain Builders

15. Processes & Practices 2 Use Algebra Lynn bought 3 party favors. Then she bought 2 more favors. Her total cost was $15. How much did each favor cost? Write a multiplication sentence with a symbol for the unknown. Solve for the unknown.

×	0	1	2	3	4	5	6	7
0	0	0	0	0	0	0	0	0
1	0	1	2	3	4	5	6	7
2	0	2	4	6	8	10	12	14
3	0	3	6	9	12	15	18	21
4	0	4	8	12	16	20	24	28

_____

16. Processes & Practices 8 Look for a Pattern Look at the multiplication table. Color the row of products of the 3s facts. Tell what pattern you see.

_____

_____

Color the column of products of the 3s facts. What can you see in the numbers in the row and column of 3s facts?

_____

17. Building on the Essential Question How can a number line help you multiply by 3?

_____

_____

368 Chapter 7 Multiplication and Division

Name

Lesson 1

Multiply by 3

Homework Helper

Need help? connectED.mcgraw-hill.com

Tyra has 3 posters on each of 3 walls in her bedroom. How many posters does Tyra have in her room? Find 3×3.

One Way Use an array to model 3 rows of 3.

 The array shows that 3 rows of 3 equals 9.

Another Way Use a number line.

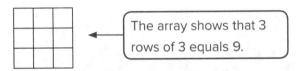

The number line shows that 3 jumps of $3 = 9$.

So, $3 \times 3 = 9$.   You can also write it like this. →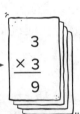

Tyra has 9 posters in her room.

Practice

Draw an array for each. Then write two multiplication sentences.

1. 3 rows of 8

2. 6 rows of 3

Lesson 1 My Homework   369

Multiply. Use the number line to skip count if needed.

0 1 2 3 4 5 6 7 8 9 10 11 12 13 14 15 16 17 18 19 20 21 22 23 24 25

3. $5 \times 3 =$ _____

4. $8 \times 3 =$ _____

5. $7 \times 3 =$ _____

6. $4 \times 3 =$ _____

Algebra Find the unknown factor. Use the Commutative Property.

7. ■ $\times 3 = 30$

   $3 \times$ ■ $= 30$

   The unknown is _____ .

8. $3 \times$ ■ $= 18$

   ■ $\times 3 = 18$

   The unknown is _____ .

Brain Builders

Write a multiplication sentence with a symbol for the unknown for Exercises 9 and 10. Then solve.

9. A box of popcorn costs $3 at the baseball game. The vendor sells 3 boxes to people in row 22, and 2 boxes to people in row 24. How much money did the vendor collect for the popcorn?

10. Gloria has a study guide for her math, social studies, and science classes. She has 3 study guides for Language Arts. Each study guide is 3 pages. How many pages of study guides does Gloria have in all?

11. **Processes &Practices** **1 Make Sense of Problems** Meredith feeds her 3 dogs twice a day. If each dog has its own bowl, how many times does she fill a dog bowl in 3 days?

12. **Test Practice** There are 3 short rows of cars in the parking lot. Each short row has 5 cars. There is one long row with 8 cars. How many cars are in the parking lot?

    Ⓐ 17 cars
    Ⓒ 23 cars
    Ⓑ 15 cars
    Ⓓ 24 cars

370 **Need more practice?** Download Extra Practice at connectED.mcgraw-hill.com

Name _____

Lesson 2
Divide by 3

ESSENTIAL QUESTION
What strategies can be used to learn multiplication and division facts?

Math in My World

Example 1

Max, Maria, and Tani have 24 markers in all. Each person has the same number of markers. How many markers does each person have?

Find the unknown quotient. $3\overline{)24}$ ← unknown quotient

One Way Use the multiplication table.

1. Locate row 3. Circle the divisor.

2. Follow row 3 to 24. Circle the dividend.

3. Move straight up the column to 8. Circle the quotient.

The unknown quotient is _____.

×	1	2	3	4	5	6	7	8
1	1	2	3	4	5	6	7	8
2	2	4	6	8	10	12	14	16
3	3	6	9	12	15	18	21	24
4	4	8	12	16	20	24	28	32
5	5	10	15	20	25	30	35	40
6	6	12	18	24	30	36	42	48
7	7	14	21	28	35	42	49	56
8	8	16	24	32	40	48	56	64

Another Way Use a related fact.

Find 24 ÷ 3 by thinking of a related multiplication fact.

Find the unknown factor. → ■ × 3 = 24

THINK What times 3 equals 24? → ____ × 3 = 24

The unknown factor is _____.

So, $3\overline{)24}$ or 24 ÷ 3 = _____. The unknown is _____.

Each person has _____ markers.

Online Content at connectED.mcgraw-hill.com

Lesson 2 371

Example 2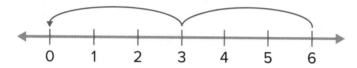

To travel to the beach, Angela and her 5 friends will divide up equally into 3 cars. How many friends will be in each car?

Find 6 ÷ 3.

Skip count backwards to find the quotient.

 Start at 6 and skip count back by 3s to 0.

 Count the number of jumps. There were _____ jumps.

So, 6 ÷ 3 = _____.

Each car will have _____ friends.

Guided Practice

Follow steps 1–3 to find 21 ÷ 3 using the multiplication table.

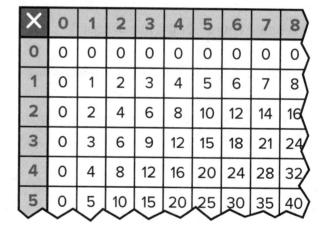

1. Locate row 3. Circle the divisor.

2. Follow row 3 to 21. Circle the dividend.

3. Circle the quotient to solve.

   21 ÷ 3 = _____

4. Skip count to find the quotient.

   12 ÷ 3 = _____

Talk MATH

Look back at the circled numbers on the multiplication table. Write the four related facts for the 3 numbers.

372 Chapter 7

Name ....

Independent Practice

Skip count backwards to find the quotient.

5.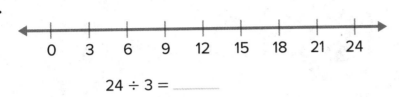

  $24 \div 3 = $ _____

Algebra Use a related multiplication fact to find the unknown.

6. $15 \div 3 = \blacksquare$

  $3 \times $ _____ $= 15$

  The unknown is _____.

7. $3 \overline{)27}$ with $\blacksquare$ on top

  $3 \times $ _____ $= 27$

  The unknown is _____.

8. $3 \overline{)21}$ with $\blacksquare$ on top

  $3 \times $ _____ $= 21$

  The unknown is _____.

Algebra Find each unknown to solve the puzzle. Write the letter corresponding to each quotient on the line above each exercise number.

9. $3 \div 3 = \blacksquare$ _____

10. $9 \div \blacksquare = 3$ _____

11. $15 \div \blacksquare = 3$ _____

12. $27 \div 9 = \blacksquare$ _____

13. $\blacksquare \div 3 = 8$ _____

14. $6 \div 2 = \blacksquare$ _____

15. $\blacksquare \div 3 = 4$ _____

16. $18 \div 3 = \blacksquare$ _____

Exercise: ___ ___ ___ ___ ___ ___ ___ ___
        9   10 11 12 13 14 15 16

Lesson 2 Divide by 3

Problem Solving

17. **Processes & Practices 4 Model Math** A soccer coach buys 3 equally-priced soccer balls for $21. What is the price for each ball? Write a division sentence. Then label each price tag.

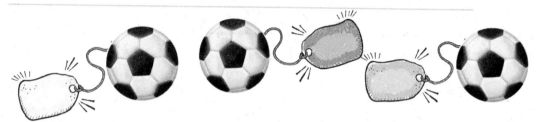

18. Stanley is on a 3-day hike. He will hike a total of 18 miles. If he hikes the same number of miles each day, how many miles will he hike the first day?

Brain Builders

19. **Processes & Practices 1 Make a Plan** Jessica arranged 27 stickers in 3 equal rows. Then she gave 1 row of stickers to Dane and 3 stickers to Kim. How many stickers does Jessica have left?

20. **Processes & Practices 7 Identify Structure** There are 24 bananas to be divided equally among 2 monkeys and a gorilla. How many bananas will each animal receive?

Rewrite the story using a related multiplication fact. Solve.

21. **Building on the Essential Question** Besides using models, how else can I find $18 \div 3$?

Name _____

Lesson 2
Divide by 3

Homework Helper

Need help? connectED.mcgraw-hill.com

Chuck and his 2 brothers read 12 books about the solar system. Each boy read the same number of books. How many books did each boy read? Find the unknown quotient.

You need to find 12 ÷ 3, or

Use the multiplication table.

1. Find row 3. Circle the divisor.

2. Move across row 3 to 12. Circle the dividend.

3. Move straight up the column to 4. Circle the quotient.

×	0	1	2	3	4	5	6
0	0	0	0	0	0	0	0
1	0	1	2	3	4	5	6
2	0	2	4	6	8	10	12
3	0	3	6	9	12	15	18
4	0	4	8	12	16	20	24
5	0	5	10	15	20	25	30

So, 12 ÷ 3 = 4. Each boy read 4 books.

Practice

Algebra Use a related multiplication fact to find the unknown.

1. 30 ÷ 3 = ■

   3 × _____ = 30

   The unknown is _____ .

2. 18 ÷ 3 = ■

   3 × _____ = 18

   The unknown is _____ .

3. 15 ÷ 3 = ■

   3 × _____ = 15

   The unknown is _____ .

4. 21 ÷ 3 = ■

   3 × _____ = 21

   The unknown is _____ .

Lesson 2 My Homework 375

Use the multiplication table to divide.

5. 24 ÷ 3 = _____

6. 9 ÷ 3 = _____

7. 27 ÷ 3 = _____

8. 3 ÷ 3 = _____

9. 18 ÷ 3 = _____

×	0	1	2	3	4	5	6	7	8	9	10
0	0	0	0	0	0	0	0	0	0	0	0
1	0	1	2	3	4	5	6	7	8	9	10
2	0	2	4	6	8	10	12	14	16	18	20
3	0	3	6	9	12	15	18	21	24	27	30
4	0	4	8	12	16	20	24	28	32	36	40
5	0	5	10	15	20	25	30	35	40	45	50
6	0	6	12	18	24	30	36	42	48	54	60
7	0	7	14	21	28	35	42	49	56	63	70
8	0	8	16	24	32	40	48	56	64	72	80
9	0	9	18	27	36	45	54	63	72	81	90
10	0	10	20	30	40	50	60	70	80	90	100

Brain Builders

10. Alana mailed 6 letters and 9 postcards in 3 different mailboxes. She put the same number of items in each mailbox. How many items did Alana put in each mailbox?

11. **Processes & Practices 2** **Use Number Sense** Ms. Banks divides 18 basketballs evenly among 3 bags. For class, she takes 2 basketballs from each bag. How many basketballs are left in one of the bags?

12. **Test Practice** Elyse served herself and 2 friends 24 ounces of juice. She poured the same amount of juice in each glass. Elyse drank 2 ounces. How many ounces of juice were left in her glass?

   Ⓐ 20 ounces   Ⓒ 8 ounces

   Ⓑ 12 ounces   Ⓓ 6 ounces

Name _____

Lesson 3
Hands On
Double a Known Fact

ESSENTIAL QUESTION
What strategies can be used to learn multiplication and division facts?

A **known fact** is a fact you have memorized. You can use a known multiplication fact to solve a multiplication fact you do not know.

Build It

Find 4 × 6.
Decompose, or separate, the number 4 into equal addends of 2 + 2.

 Model the known fact, 2 × 6.
Use counters to make an array.
Show 2 rows of 6. Draw your array.

Write the number sentence.

_____ × _____ = _____

 Double the known fact.
Make one more 2 × 6 array. Draw your array.
Write the number sentence for this array.

_____ × _____ = _____

Add the two products together.

12 + _____ = _____

 Find 4 × 6.
Push the two arrays together. Write the new number sentence.

_____ × _____ = _____

2 × 6 plus 2 × 6 = _____   **So,** 4 × 6 = _____

Online Content at connectED.mcgraw-hill.com

Try It

Find 6 × 5.

Decompose 6 into two equal addends. 6 = 3 + _____

 Model the known fact, 3 × 5 two times.

Draw two 3 × 5 arrays.
Write the product on each array.

 Double the known fact 3 × 5.

Draw the two arrays pushed together.

Add the products.

15 + _____ = _____

 Find 6 × 5.

Write the multiplication sentence.

_____ × _____ = _____

3 × 5 plus 3 × 5 = _____ . So, 6 × 5 = _____ .

Talk About It

1. Why can you double the product of 2 × 6 to find 4 × 6?

_____

_____

2. What fact would doubling 3 × 6 help you find? _____

3. Give an example of doubling a known fact. Explain.

_____

_____

4. Draw two arrays that you could put together to find 4 × 5. Label the arrays with a number sentence.

Name ....

Practice It

Use counters to model a known fact that will help you find the first product. Draw the model two times.

5. 4 × 3 = _____

Known fact: 2 × 3 = _____

Double the product: 6 + 6 = _____

6. 4 × 4 = _____

Known fact: 2 × 4 = _____

Double the product: 8 + 8 = _____

7. 7 × 4 = _____

Known fact: _____

Double the product: _____

8. 6 × 7 = _____

Known fact: _____

Double the product: _____

Use counters to double the known fact. Write the product it helps you find.

9. 3 × 8 = _____

3 × 8 = _____

_____ × _____ = _____

10. 3 × 6 = _____

3 × 6 = _____

_____ × _____ = _____

Lesson 3 Hands On: Double a Known Fact 379

Apply It

11. Megan and Paul each have a tray of cookies. Each tray has 2 rows of 6 cookies. How many cookies do they have if they put their trays together?

What known fact did you double? _____

What fact is found by doubling the known fact?

_____

12. Processes &Practices **4** **Model Math** Two heart buttons have 8 holes. Jeffery doubles the number of heart buttons shown. How many holes will there be now?

What known fact did you double? _____

What fact is found by doubling the known fact?

_____

13. Processes &Practices **3** **Justify Conclusions** Can you double a known fact to find 7 × 6? Explain.

_____

_____

Write About It

14. When is doubling a known fact useful?

_____

_____

380 Chapter 7 Multiplication and Division

Name

MY Homework

Lesson 3

Hands On: Double a Known Fact

Homework Helper

Need help? connectED.mcgraw-hill.com

Find 6 × 5.

Decompose, or separate, the factor 6 into two equal addends of 3 + 3. Then you can double the known fact, 3 × 5.

1. **Model the known fact, 3 × 5.**
   Use counters to make an array that shows 3 rows of 5.

   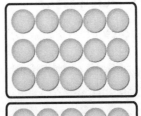
   3 × 5 = 15

2. **Double the known fact.**
   Make another array that shows 3 rows of 5.

   3 × 5 = 15

3. **Find 6 × 5.**
   Push the two arrays together into 6 rows of 5.
   Add the products from the two arrays:

   15 + 15 = 30

   The combined arrays show 6 × 5 = 30.

   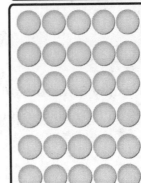
   6 × 5 = 30

Practice

1. Draw counters to model a known fact that will help you find 4 × 5. Draw the model two times.

   4 × 5 = _____

   Known fact: 2 × 5 = _____

   Double the product: 10 + 10 = _____

Double the known fact. Write the product it helps you find.

2. $3 \times 7 =$ _____

   $3 \times 7 =$ _____

   _____ $\times$ _____ $=$ _____

3. $3 \times 3 =$ _____

   $3 \times 3 =$ _____

   _____ $\times$ _____ $=$ _____

4. $2 \times 6 =$ _____

   $2 \times 6 =$ _____

   _____ $\times$ _____ $=$ _____

5. $2 \times 9 =$ _____

   $2 \times 9 =$ _____

   _____ $\times$ _____ $=$ _____

Problem Solving

Double a known fact to solve.

6. Dr. Berry sees 3 patients every hour. If she works 8 hours, how many patients does she see?

   _____

7. The Johnson twins and the Clayton twins went to the fair. Each child rode 5 rides. What is the total number of times all four children went on a ride?

   _____

8. **Processes & Practices** 7 **Identify Structure** Vince drinks 4 large glasses of water each day. How many glasses of water does Vince drink in 7 days?

   _____

Vocabulary Check

Choose the correct word(s) to complete each sentence.

known fact   decompose

9. One way to _____ the number 8 is to write it as $4 + 4$.

10. A fact that you have memorized is a _____ .

Name _____

Lesson 4
Multiply by 4

ESSENTIAL QUESTION
What strategies can be used to learn multiplication and division facts?

Math in My World

Example 1

A box is packed with 4 rows of oranges. Each row has 9 oranges. How many oranges are in the box?

Find 4 × 9.

Decompose 4 into equal addends of 2 + 2.
Use the **known fact** of 2 × 9 and double it.

4 × 9 = 2 × 9 + _____ × _____   Multiply.

= _____ + 18   Add.

= 36

The array shows that 2 × 9 plus _____ × 9 equals _____ × 9.

So, 4 × 9 = _____. There are _____ oranges in the box.

Example 2

There are 3 bunches of bananas. Each bunch has 4 bananas. How many bananas are there all together? Write a multiplication sentence with a symbol for the unknown. Then solve.

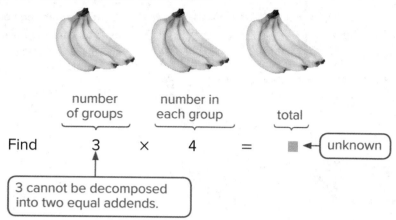

Find   3   ×   4   =   ■ ← unknown

(number of groups) (number in each group) (total)

3 cannot be decomposed into two equal addends.

Solve by decomposing the factor 4 into two equal addends of 2. Use the known fact 3 × 2 plus 3 × 2 to find the unknown.

3 × 2 plus _____ × _____ = ■

6 + _____ = _____

So, 3 × 4 = _____. The unknown is _____.

There are _____ bananas all together.

Guided Practice

1. Double a known fact to find 6 × 4. Label the array and complete the number sentences.

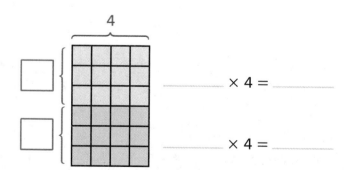

_____ × 4 = _____

_____ × 4 = _____

12 + 12 = _____

So, 6 × 4 = _____.

Talk MATH

Explain how knowing 2 × 7 can help you find 4 × 7.

384 Chapter 7 Multiplication and Division

Name ....

Independent Practice

Double a known fact to find each product. Draw and label an array.

2. $8 \times 4 =$ _____

3. $5 \times 4 =$ _____

4. $4 \times 6 =$ _____

5. $7 \times 4 =$ _____

Algebra Find each unknown. Double a known fact.

6. $7 \times 4 = \blacksquare$

The unknown is _____ .

7. $9 \times 4 = \blacksquare$

The unknown is _____ .

8. $\begin{array}{r} 4 \\ \times\ 4 \\ \hline \blacksquare \end{array}$

The unknown is _____ .

9. $\begin{array}{r} 4 \\ \times\ 10 \\ \hline \blacksquare \end{array}$

The unknown is _____ .

Lesson 4 Multiply by 4   **385**

Problem Solving

10. The library offers 4 activities on Saturday. Each activity table has room for 10 children. How many children can take part in the activities? Write a multiplication sentence to solve.

_____

11. **Processes & Practices 2** **Reason** Mitch bought 4 bottles of suntan lotion for $6 each. Later the lotion went on sale for $4 each. How many more bottles could he have bought for the same amount of money if he waited for the sale?

_____

Brain Builders

12. **Processes & Practices 2** **Use Number Sense** Look at the multiplication table. Circle the two numbers that represent the product of 4 and 10. Write this product as a sum of two equal addends.

×	0	1	2	3	4	5	6	7	8	9	10
0	0	0	0	0	0	0	0	0	0	0	0
1	0	1	2	3	4	5	6	7	8	9	10
2	0	2	4	6	8	10	12	14	16	18	20
3	0	3	6	9	12	15	18	21	24	27	30
4	0	4	8	12	16	20	24	28	32	36	40
5	0	5	10	15	20	25	30	35	40	45	50
6	0	6	12	18	24	30	36	42	48	54	60
7	0	7	14	21	28	35	42	49	56	63	70
8	0	8	16	24	32	40	48	56	64	72	80
9	0	9	18	27	36	45	54	63	72	81	90
10	0	10	20	30	40	50	60	70	80	90	100

Can the product of 4 and any number always be written as a sum of two equal addends? Explain.

_____

_____

13. **Building on the Essential Question** What is one strategy I could use to multiply by 4? For what other numbers could I use a similar strategy? Explain.

_____

_____

386 Chapter 7 Multiplication and Division

Name ....

MY Homework

Lesson 4

Multiply by 4

Homework Helper

Need help? connectED.mcgraw-hill.com

Find 4 × 7.

Decompose 4 into equal addends of 2 + 2.

$4 \times 7 = 2 \times 7 + 2 \times 7$
$= 14 + 14$
$= 28$

The array shows that 4 groups of 7 equals 2 groups of 7 plus 2 groups of 7.

So, 4 × 7 = 28.

Practice

Double a known fact to find the product. Label the array and complete the number sentences.

1. 4 × 5

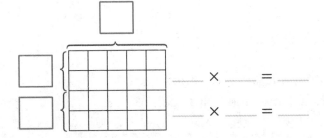

____ × ____ = ____

____ × ____ = ____

____ + ____ = ____

So, 4 × 5 = ____.

2. 3 × 4

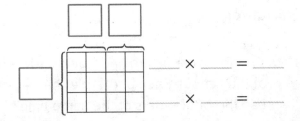

____ × ____ = ____

____ × ____ = ____

____ + ____ = ____

So, 3 × 4 = ____.

Lesson 4 My Homework 387

Double a known fact to find each product. Draw and label an array.

3. 6 × 4 = _____

4. 4 × 8 = _____

Algebra Find each unknown. Double a known fact.

5. 9 × 4 = ■

The unknown is _____.

6. 4 × 4 = ■

The unknown is _____.

Brain Builders

7. **Processes &Practices** **Model Math** Melissa owns 4 sets of jewelry. Each set contains 2 earrings, 1 necklace, 1 bracelet, and 1 ring. How many pieces of jewelry does Melissa have all together? Write a multiplication sentence. Explain how you can double a known fact to find the answer.

_____

_____

Vocabulary Check

Write a definition for the vocabulary word(s).

8. known fact _____

9. decompose _____

10. **Test Practice** There are 4 birdfeeders at the park and 3 birdfeeders at the school. Each birdfeeder has 4 perches. How many birds can use the birdfeeders at the same time?

Ⓐ 32 birds   Ⓒ 11 birds

Ⓑ 28 birds   Ⓓ 3 birds

Name _____

Lesson 5
Divide by 4

ESSENTIAL QUESTION
What strategies can be used to learn multiplication and division facts?

Math in My World

Example 1

The distance around a square window in Peter's house is **12 feet**. What is the length of each side?

Find 12 ÷ 4.

One Way Use models.

Start with 12 counters. Circle 4 equal groups.

There are _____ counters in each group. 12 ÷ 4 = _____

So, the length of each side of the window is _____ feet.

Another Way Use repeated subtraction.

Subtract groups of 4 until you reach 0.

Count the number of times you subtracted.

Groups of _____ were subtracted _____ times.

So, 12 ÷ 4 = _____.

①
```
  12
-  4
────
   8
```

②
```
  ☐
-  4
────
  ☐
```

③
```
  ☐
-  4
────
   0
```

Online Content at connectED.mcgraw-hill.com

Lesson 5 **389**

Example 2

An ostrich egg weighs 4 pounds. The total weight of the eggs in a nest is 28 pounds. How many ostrich eggs are there?

Find the unknown in 28 ÷ 4 = ■ or 4)28.

Draw an array. Then use the inverse operation, multiplication, to find the unknown.

Think: ■ × 4 = 28. ← unknown factor

You know that _____ × 4 = 28.

So, then 28 ÷ 4 = _____ or 4)28.

The unknown is _____. There are _____ ostrich eggs.

> **Helpful Hint**
> Multiplication is the inverse operation of division.

Guided Practice

Use counters to find how many are in each group.

1. 8 counters
   4 equal groups

   _____ in each group

   So, 8 ÷ 4 = _____.

2. 24 counters
   4 equal groups

   _____ in each group

   So, 24 ÷ 4 = _____.

3. Use repeated subtraction to find 20 ÷ 4.

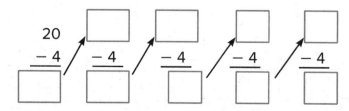

So, 20 ÷ 4 = _____.

> **Talk MATH**
> Without dividing, how do you know that the quotient of 12 ÷ 3 is greater than the quotient of 12 ÷ 4?

390 Chapter 7 Multiplication and Division

Name _____

Independent Practice

Use counters to find the number of equal groups or how many are in each group.

4. 28 counters
   4 equal groups

   _____ in each group

   So, 28 ÷ 4 = _____, or $4\overline{)28}$

5. 4 counters
   4 equal groups

   _____ in each group

   So, _____ ÷ _____ = _____ or $4\overline{)4}$

Use repeated subtraction to divide.

6. 8 ÷ 4 = _____

7. 16 ÷ 4 = _____

Algebra Draw an array and use the inverse operation to find each unknown.

8. ? × 6 = 24

   ■ ÷ 4 = 6

   ? = _____

   ■ = _____

9. 8 × ? = 32

   32 ÷ 4 = ■

   ? = _____

   ■ = _____

Lesson 5 Divide by 4 391

Problem Solving

Algebra Write a division sentence with a symbol for the unknown in Exercises 10 and 11. Solve.

10. Gwen, Clark, Elvio, and Trent will be on vacation for 20 days. They are dividing the planning equally. How many days will Clark have to plan?

_____

11. A bus has 32 pieces of luggage. If each person brought 4 pieces of luggage, how many people are on the trip?

_____

Brain Builders

12. **Processes &Practices 2** **Reason** It costs $40 for 4 friends to ride go-carts for 1 hour. For how many hours can 1 person ride for $20?

_____

13. **Processes &Practices 3** **Find the Error** Kendra wrote the two number sentences below to help her find 12 ÷ 4. Explain and correct her mistake.

$$4 + 8 = 12$$
$$\text{So, } 12 \div 4 = 8.$$

_____
_____
_____

14. **Building on the Essential Question** How can an array help me divide?

_____
_____
_____

Name ....

MY Homework

Lesson 5

Divide by 4

Homework Helper

Need help? connectED.mcgraw-hill.com

The Dolan family bought a pack of 20 juice boxes. There are 4 people in the Dolan family. If they divide the juice boxes evenly, how many will each family member get?

Find the unknown in $20 \div 4 = \blacksquare$, or $4\overline{)20}$.

Use multiplication to find the unknown.

Think: $\blacksquare \times 4 = 20$

You know that $5 \times 4 = 20$.

So, $20 \div 4 = 5$, or $4\overline{)20}^{\,5}$.

The unknown is 5. Each family member will get 5 juice boxes.

Practice

Use counters to find how many are in each group.

1. 8 counters
   4 equal groups

   _____ in each group

   So, $8 \div 4 = $ _____.

2. 40 counters
   4 equal groups

   _____ in each group

   So, $40 \div 4 = $ _____.

3. 28 counters
   4 equal groups

   _____ in each group

   So, $28 \div 4 = $ _____.

4. 12 counters
   4 equal groups

   _____ in each group

   So, $12 \div 4 = $ _____.

Lesson 5 My Homework 393

Use repeated subtraction to divide.

5. $24 \div 4 =$ _____

$$\begin{array}{r} 24 \\ -4 \\ \hline 20 \end{array}$$

6. $16 \div 4 =$ _____

Algebra Use the inverse operation to find each unknown.

7. _____ $\times 2 = 8$

_____ $\div 4 = 2$

8. $9 \times$ _____ $= 36$

$36 \div 4 =$ _____

Brain Builders

Write a division sentence to solve.

9. A boat rental shop has enough red boats for 20 people and blue boats for 16 people to ride. Each boat seats 4 people. How many boats does the shop have?

_____

10. **Processes & Practices 6** **Explain to a Friend** Ollie shared 24 marbles evenly with 3 friends. Kaitlyn shared 18 marbles evenly with 2 friends. Did Ollie or Kaitlyn's friends get more marbles? Explain.

_____

11. **Test Practice** Each minute, 4 gallons of water flow into a tub that holds 40 gallons. There are 32 gallons of water in the tub. How much longer can the water run before it overflows?

Ⓐ 2 minutes   Ⓒ 8 minutes

Ⓑ 6 minutes   Ⓓ 10 minutes

Check My Progress

Vocabulary Check

Label each with the correct word(s).

decompose   dividend   divisor

inverse operations   **known fact**   quotient

1. _____

7 × 4 = 28   28 ÷ 7 = 4

2. _____   3. _____   4. _____

5. A _____ is a fact you have memorized.

6. You can _____ 6 into two equal addends; 3 + 3.

Concept Check

Draw an array for each. Then write two multiplication sentences.

7. 3 rows of 6

8. 4 rows of 6

9. Draw jumps on the number line to find 27 ÷ 3.

27 ÷ 3 = _____

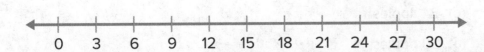

Algebra Use a related multiplication fact to find the unknown.

10. $21 \div 3 = \blacksquare$

    $\blacksquare \times 3 = 21$

    The unknown is _____.

11. $32 \div 4 = \blacksquare$

    $4 \times \blacksquare = 32$

    The unknown is _____.

Double a known fact to find each product. Draw an array.

12. $6 \times 5 =$ _____

13. $6 \times 7 =$ _____

Brain Builders

14. There are 4 quarts in one gallon of milk. Lawrence has 5 quarts of milk in the refrigerator. How many more quarts of milk are in the gallons shown below than are in the refrigerator?

_____

15. **Test Practice** Aisha picked 8 green apples, 9 yellow apples, and 10 red apples. She placed an equal number of apples in 3 bags. How many apples did she place in each bag?

    Ⓐ 8 apples   Ⓒ 24 apples

    Ⓑ 9 apples   Ⓓ 30 apples

Name ....

Lesson 6
Problem-Solving Investigation
STRATEGY: Extra or Missing Information

ESSENTIAL QUESTION
What strategies can be used to learn multiplication and division facts?

Learn the Strategy

The school's hayride starts at 6 P.M. There are 4 wagons that can hold 9 children each. Half of the children going are girls. What is the total number of children that can ride on the wagons?

1 Understand
What facts do you know?

The hayride starts at _____ P.M.

There are _____ wagons that hold _____ children each.

Half of the children are girls.

What do you need to find?

the number of _____ that can ride on the 4 wagons

2 Plan
Decide what facts are important.

the number of _____

the number of _____ each wagon holds

Extra Information
- The hayride starts at 6 P.M.
- Half of the children are girls.

3 Solve

$\underline{4} \times \underline{9} = \underline{}$
wagons children   total children

4 Check
Does your answer make sense? Explain.

Online Content at connectED.mcgraw-hill.com

Lesson 6 397

Practice the Strategy

An animal shelter has 23 cats and 14 dogs. How much would it cost to adopt 1 cat and 1 dog? Tell what facts are extra or missing.

1 Understand

What facts do you know?

_____

_____

What do you need to find?

_____

2 Plan

_____

_____

_____

_____

3 Solve

4 Check

Does your answer make sense? Explain.

_____

_____

Name _____

Apply the Strategy

Determine if there is extra or missing information to solve each problem. Then solve if possible.

1. **Processes & Practices 3**  **Draw a Conclusion** Mrs. Friedman had 2 boxes of chalk. She bought 4 more boxes with 10 pieces each. She paid $2 per box. How much did she spend on the 4 boxes of chalk?

   _____

   _____

Brain Builders

2. Bert bought 4 of each of the items below. How much change did he get back?

Item	Cost
Pencils	$2
Paper	$1
Binder	$3

   What missing information do you need to solve the problem?

   _____

3. **Processes & Practices 5** **Use Math Tools** Alejandra is 58 inches tall. Her sister is in the first grade and is 10 inches shorter than Alejandra. What known fact when doubled, is equal to the height of Alejandra's sister?

   _____

   _____

   Alejandra's cousin is 22 inches shorter than her. What known fact, when doubled, is equal to the height of Alejandra's cousin?

   _____

Lesson 6 Problem-Solving Investigation 399

Review the Strategies

Use any strategy to solve each problem.
- Determine extra or missing information.
- Make a table.
- Look for a pattern.
- Use models.

4. Ten of Eduardo's baseball cards are All-Star cards. His friend has four times as many All-Star cards. How many All-Star baseball cards does his friend have?

_____

5. The third grade class had four chicks hatch every day for 5 days. Nine of the chicks were yellow, and the rest were brown. How many chicks hatched in all?

_____

_____

6. Four friends bought a total of 24 computer games. Each friend bought the same number of games. How many games did each friend buy?

_____

7. Processes & Practices 1 Make Sense of Problems There are 4 sheets of stickers. Each sheet has 7 stickers. Kyla gives 1 sheet of stickers to her friend. How many stickers does Kyla have left?

_____

8. How many dots will there be all together if there are 2 dominoes like the one below?

_____

400 Chapter 7 Multiplication and Division

Name _____

MY Homework

Lesson 6

Problem Solving: Extra or Missing Information

Homework Helper

Need help? connectED.mcgraw-hill.com

Arthur will cut a 12-foot board into 4 equal pieces. If it takes him 10 minutes to cut the board how long will each piece be?

1 Understand

What facts do you know?
Arthur has a 12-foot long board.
He will cut the board into 4 equal pieces.
It will take 10 minutes to cut the board.

What do you need to find?
the length of the 4 pieces

2 Plan

Decide which facts are important.
- the length of the board being cut
- the number of pieces being cut

Extra Information
It will take him 10 minutes to cut the board.

3 Solve

Divide the board's length by the number of pieces being cut.
$12 \div 4 = 3$.
So, each piece will be 3 feet long.

4 Check

Does the answer make sense?
Use the inverse operation, multiplication, to check.
$3 \times 4 = 12$
So, the answer is reasonable.

Lesson 6 My Homework

Problem Solving

Determine if there is extra or missing information to solve each problem. Then solve if possible.

1. Brandy ate 9 corn chips for a snack. She ate twice as many raisins. She also drank a juice box. How many raisins did Brandy eat?

   _____

   _____

2. Miguel bought a movie ticket that cost $5. He also bought popcorn and a bottle of water. How much did Miguel spend all together?

   _____

Brain Builders

3. Annie has basketball practice from 2:30 P.M. to 4:00 P.M. Every practice, between 2:30 and 3:30, she makes 6 free throws. Between 3:30 and 4:00, she makes 4 free throws. How many free throws does she make in 4 practices?

   _____

   _____

4. Naya has 12 flowers. She gives away 6 to Jane and 3 to Heather. Hannah has 5 flowers. How many flowers does Naya have left?

   _____

   _____

5. **Processes & Practices 3** **Draw a Conclusion** Juan bought two new tires for his bike. He also had the bike tuned up. The total bill was $65. How much did Juan spend on the 2 tires?

   _____

   _____

Lesson 7
Multiply by 0 and 1

ESSENTIAL QUESTION
What strategies can be used to learn multiplication and division facts?

Math in My World

Example 1

There are 4 flower pots. Each has 1 daisy. How many daisies are there in all? Find the unknown.

$4 \times 1 = \blacksquare$ ← unknown

Circle 4 equal groups of 1.

1 + 1 + 1 + 1 = _____

So, 4 groups of _____ is _____. The unknown is _____.

There are _____ daisies in all.

The **Identity Property of Multiplication** says that when any number is multiplied by _____, the product is that number.

Online Content at connectED.mcgraw-hill.com

Lesson 7 403

Example 2

1 Describe the pattern of the products shaded green.

Each product is the number _____.

×	0	1	2	3	4	5	6	7	8	9	10
0	0	0	0	0	0	0	0	0	0	0	0
1	0	1	2	3	4	5	6	7	8	9	10
2	0	2	4	6	8	10	12	14	16	18	20
3	0	3	6	9	12	15	18	21	24	27	30
4	0	4	8	12	16	20	24	28	32	36	40
5	0	5	10	15	20	25	30	35	40	45	50
6	0	6	12	18	24	30	36	42	48	54	60
7	0	7	14	21	28	35	42	49	56	63	70
8	0	8	16	24	32	40	48	56	64	72	80
9	0	9	18	27	36	45	54	63	72	81	90
10	0	10	20	30	40	50	60	70	80	90	100

2 Look at the circled product. Follow left and above to the circled factors. Complete the number sentence.

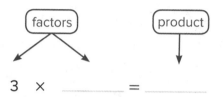

3 × _____ = _____

3 Look at the product with a square around it. Complete the number sentence.

7 × _____ = _____

> If there are 7 equal groups and 0 objects in each group, then there are 0 objects total.

The two number sentences are examples of the **Zero Property of Multiplication** which states that the product of any number and zero is zero.

Guided Practice

Look at the row and column of products shaded yellow. Describe the pattern.

×	0	1	2	3	4	5
0	0	0	0	0	0	0
1	0	1	2	3	4	5
2	0	2	4	6	8	10
3	0	3	6	9	12	15
4	0	4	8	12	16	20
5	0	5	10	15	20	25

1. When a number is multiplied by _____, the product is that number.

2. This is an example of the _____ Property of Multiplication.

Talk MATH
If 100 is multiplied by 0, what will be the product? Explain your reasoning.

Name

Independent Practice

Algebra Circle equal groups to find the unknown. Write the unknown.

3. $2 \times 1 = \blacksquare$ or $\begin{array}{r} 2 \\ \times\ 1 \\ \hline \blacksquare \end{array}$

The unknown is _____.

4. $5 \times 1 = \blacksquare$ or $\begin{array}{r} 5 \\ \times\ 1 \\ \hline \blacksquare \end{array}$

The unknown is _____.

Write an addition sentence to help find each product. Check using the Identity Property of Multiplication.

5. $8 \times 1 = $ _____

6. $7 \times 1 = $ _____

Use the Zero Property of Multiplication to find each product.

7. $10 \times 0 = $ _____

8. $6 \times 0 = $ _____

9. $3 \times 0 = $ _____

10. **Algebra** Draw a line to match each number sentence to its unknown.

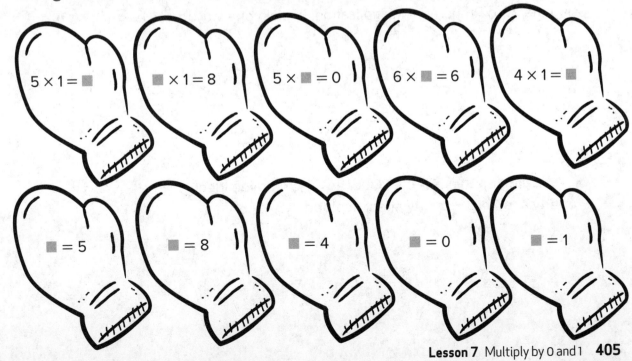

Lesson 7 Multiply by 0 and 1   405

Problem Solving

Algebra Write a multiplication sentence with a symbol for the unknown. Then solve.

11. There is 1 student sitting at each of the 9 tables in the library. How many students are there all together?

_____

12. **Processes &Practices 4** **Model Math** How many legs do 8 snakes have?

_____

Brain Builders

13. How many puppies are there all together if there is 1 wagon with 27 puppies and 35 kittens? Explain your reasoning.

_____
_____

14. **Processes &Practices 7** **Identify Structure** Write a real-world problem using the Zero Property of Multiplication. Explain one way to solve.

_____
_____
_____

15. **Building on the Essential Question** How do the Identity Property and Zero Property affect numbers?

_____
_____
_____

406 Chapter 7 Multiplication and Division

Name

Lesson 7

Multiply by 0 and 1

Homework Helper

Need help? connectED.mcgraw-hill.com

Three customers each bought a cone with 1 scoop of frozen yogurt. How many scoops did the customers buy in all?

Find 3 × 1.

The Identity Property of Multiplication states that when any number is multiplied by 1, the product is that number.
So, 3 × 1 = 3.

The 3 customers have eaten their frozen yogurt cones. Now how many scoops do the customers have in all?

Find 3 × 0.

The Zero Property of Multiplication states that when any number is multiplied by 0, the product is 0.
So, 3 × 0 = 0.

Practice

Algebra Circle equal groups to find the unknown. Write the unknown.

1. 8 × 1 = ▪

   The unknown is _____ .

2. 5 × 1 = ▪

   The unknown is _____ .

Lesson 7 My Homework 407

Use the Identity Property of Multiplication or the Zero Property of Multiplication to find each product.

3. $4 \times 0 =$ _____
4. $7 \times 1 =$ _____
5. $7 \times 0 =$ _____
6. $6 \times 1 =$ _____
7. $1 \times 0 =$ _____
8. $9 \times 1 =$ _____
9. $2 \times 1 =$ _____
10. $8 \times 1 =$ _____
11. $5 \times 0 =$ _____

Brain Builders

Algebra Write a multiplication sentence with a symbol for the unknown for Exercises 12 and 13. Then solve.

12. Jordan collects stamps. If he gets 1 stamp a day for 7 days one week and 5 days the next week, how many stamps will he add to his collection?

13. **Processes &Practices** **Be Precise** There are no pockets on any of Henry's shirts or jackets. How many pockets do 2 of Henry's shirts and 4 of Henry's jackets have?

Vocabulary Check

Write the correct word to complete the sentence.

    Identity   Zero

14. The _____ Property of Multiplication states that any number multiplied by 0 equals 0.

15. The _____ Property of Multiplication states that any number multiplied by 1 is that number.

16. **Test Practice** Which multiplication sentence shows how to find the number of wings two alligators have all together?

 Ⓐ $1 + 1 = 2$ Ⓒ $1 \times 1 = 1$

 Ⓑ $2 \times 2 = 4$ Ⓓ $2 \times 0 = 0$

Lesson 8
Divide with 0 and 1

ESSENTIAL QUESTION
What strategies can be used to learn multiplication and division facts?

There are rules you can use when you divide with 0 or 1.

Math in My World

Example 1

There are 3 friends and only 1 bean bag chair. How many friends will be on the bean bag chair?

Find 3 ÷ 1. Use Xs to draw the friends on the chair.

So, 3 ÷ 1 = _____.

There will be _____ friends on the bean bag chair.

Helpful Hint
When a number is divided by 1, the quotient is that number.

Example 2

There are 3 friends and 3 bean bag chairs. How many friends will be on each bean bag chair?

Find 3 ÷ 3.
Use Xs to equally partition the friends on the chairs.

So, 3 ÷ 3 = _____.

There will be _____ friend on each bean bag chair.

Helpful Hint
When a number is divided by itself, the quotient is 1.

Online Content at connectED.mcgraw-hill.com

Example 3

There are 0 friends and 3 bean bag chairs. How many friends will sit on each bean bag chair?

Find 0 ÷ 3. Are there friends to draw? _____

So, 0 ÷ 3 = _____. There will be _____ friends on each bean bag chair.

Helpful Hint
When zero is divided by a nonzero number, the quotient is zero.

Example 4

There are 3 friends and 0 bean bag chairs. How many friends will sit on each bean bag chair?

Find 3 ÷ 0.

So, the quotient of 3 ÷ _____ cannot be found.

Helpful Hint
It is impossible to divide a number by zero.

Guided Practice

Draw models to find each quotient.

1. 6 ÷ 1 = _____   or   ☐ 1)6

2. 4 ÷ 4 = _____   or   ☐ 4)4

Talk MATH
How do you know you can divide any number by 1 or itself?

410 Chapter 7 Multiplication and Division

Name _____

Independent Practice

Find each quotient. Draw lines to match each division sentence to its model.

3. 2)‾2 □

4. 1 ÷ 1 = ____

5. 5)‾0 □

6. 5 ÷ 1 = ____

7. 1)‾4 □

8. 0 ÷ 2 = ____

9. 4)‾4 □

10. 0 ÷ 1 = ____

Write a division sentence as an example of each division rule.

11. When a number is divided by itself, the quotient is 1. _____

12. When 0 is divided by a nonzero number, the quotient is 0. _____

13. When a number is divided by 1, the quotient is that number. _____

Lesson 8 Divide with 0 and 1   411

Problem Solving

Write a division sentence to solve.

14. There are 7 students and one table. If the same number of students must sit at each table, how many students will sit at each table?

15. **Processes &Practices 4** **Model Math** Mia and her 4 friends equally divide up 5 glasses of juice. How many glasses of juice will each friend get?

Brain Builders

16. There are no dogs to sleep in the dog beds. There is 1 cat. How many dogs will sleep in each dog bed?

17. **Processes &Practices 1** **Make Sense of Problems** There are 17 students in Mr. Macy's class and 18 students in Mrs. Marshall's class. To play a game, each person needs 1 playing piece. How many playing pieces are needed for the class to play the game? Write a division sentence to solve.

18. **Building on the Essential Question** How do division rules help me learn division facts more quickly?

412 Chapter 7 Multiplication and Division

Name _____

MY Homework

Lesson 8

Divide with 0 and 1

Homework Helper

Need help? connectED.mcgraw-hill.com

Larry has 3 keys. He divides them evenly among 1 keychain. How many keys are on each keychain?

Find 3 ÷ 1.

Use counters to model.

Divide 3 counters evenly into 1 group.

So, 3 ÷ 1 = 3.

There are 3 keys on the keychain.

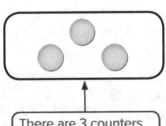
There are 3 counters in one group.

Practice

Complete the division sentence for each model.

1.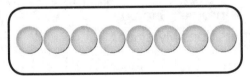

    8 ÷ 1 = _____

2.

    0 ÷ 2 = _____

3.

    0 ÷ 4 = _____

4.

    1 ÷ 1 = _____

Lesson 8 My Homework 413

Algebra Use a related multiplication fact to find the unknown.

5. 9 ÷ 9 = ■

   9 × _____ = 9

   The unknown is _____.

6. 0 ÷ 6 = ■

   6 × _____ = 0

   The unknown is _____.

7. 0 ÷ 8 = ■

   8 × _____ = 0

   The unknown is _____.

8. 2 ÷ 2 = ■

   2 × _____ = 2

   The unknown is _____.

Brain Builders

Write a division sentence to solve.

9. There are 9 boys and 6 girls who want to share 15 apples. How many apples will each child get?

   _____

10. **Processes & Practices** 7 **Identify Structure** Mrs. Perkins needs 24 sheets of red paper so she can give one to the 11 boys and 13 girls in her class. She looked on the shelf, and there are no sheets of red paper left. How many sheets of red paper can Mrs. Perkins hand out?

    _____

11. Lionel bought 3 model rockets. He shares them equally between himself and 2 friends. How many rockets does each boy have?

    _____

12. Myra draws 3 elephants and 2 giraffes for a class project. She puts each drawing in a separate folder. How many folders does Myra use?

    _____

13. **Test Practice** Angelina has 4 fiction and 2 nonfiction books. She has 1 backpack to carry her books. How many books does Angelina have in her backpack?

    Ⓐ 7 books   Ⓒ 1 book

    Ⓑ 6 books   Ⓓ 0 books

Name _____

Fluency Practice

Processes & Practices 6

Multiply.

1. $4 \times 9 =$ _____
2. $5 \times 3 =$ _____
3. $4 \times 6 =$ _____

4. $3 \times 6 =$ _____
5. $3 \times 2 =$ _____
6. $4 \times 4 =$ _____

7. $2 \times 2 =$ _____
8. $0 \times 7 =$ _____
9. $4 \times 5 =$ _____

10. $2 \times 5 =$ _____
11. $3 \times 7 =$ _____
12. $1 \times 2 =$ _____

13.   1
    $\times 3$

14.   4
    $\times 3$

15.   3
    $\times 8$

16.   0
    $\times 4$

17.   4
    $\times 7$

18.   3
    $\times 9$

19.   4
    $\times 8$

20.   1
    $\times 8$

21.   4
    $\times 2$

22.   3
    $\times 1$

23.   0
    $\times 9$

24.   1
    $\times 5$

Online Content at connectED.mcgraw-hill.com

Name _____

Fluency Practice

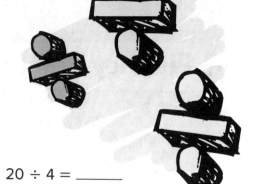

Divide.

1. 27 ÷ 3 = _____
2. 21 ÷ 3 = _____
3. 20 ÷ 4 = _____

4. 8 ÷ 4 = _____
5. 16 ÷ 4 = _____
6. 24 ÷ 3 = _____

7. 32 ÷ 4 = _____
8. 9 ÷ 3 = _____
9. 7 ÷ 1 = _____

10. 0 ÷ 9 = _____
11. 18 ÷ 3 = _____
12. 6 ÷ 1 = _____

13. 3)12
14. 4)28
15. 2)0
16. 3)15

17. 2)2
18. 3)6
19. 1)9
20. 4)24

21. 1)4
22. 3)30
23. 4)36
24. 8)0

416 Chapter 7 Multiplication and Division

Review

Chapter 7
Multiplication and Division

Vocabulary Check

Use the word bank below to complete each clue.

decompose   Identity Property of Multiplication

inverse operations   known fact   Zero Property of Multiplication

1. The _____ says that when any number is multiplied by 1, the product is that number.

2. A _____ is a fact you have memorized.

3. Multiplication and division are _____ because they undo each other.

4. _____ means to take a number apart.

5. The _____ says that when you multiply a number by 0, the product is zero.

Concept Check

Algebra Use a related multiplication fact to find the unknown.

6. 16 ÷ 4 = ■

   4 × ■ = 16

   The unknown is ____.

7. 24 ÷ 3 = ■

   3 × ■ = 24

   The unknown is ____.

8. 20 ÷ 4 = ■

   4 × ■ = 20

   The unknown is ____.

My Chapter Review 417

Use counters to model a known fact that will help you find the first product. Draw the model two times.

9. 6 × 6 = _____

    Known fact:

    _____ × _____ = _____

    Double the product:

    _____ + _____ = _____

10. 7 × 4 = _____

    Known fact:

    _____ × _____ = _____

    Double the product:

    _____ + _____ = _____

Algebra Find each unknown. Double a known fact.

11. 7 × 6 = ■

    The unknown is _____ .

12. 9 × 4 = ■

    The unknown is _____ .

Write an addition sentence to help find each product.

13. 6 × 1 = _____

    1 + 1 + _____ + _____ + _____ + _____ = _____

14. 7 × 1 = _____

    1 + 1 + 1 + _____ + _____ + _____ + _____ = _____

Use the Zero Property of Multiplication to find each product.

15. 7 × 0 = _____

16. 9 × 0 = _____

17. 6 × 0 = _____

Write a division sentence as an example of each division rule.

18. When a number is divided by 1, the quotient is that number.

19. When a number is divided by itself, the quotient is 1.

20. When 0 is divided by a nonzero number, the quotient is 0.

418 Chapter 7 Multiplication and Division

Problem Solving

21. There are a total of 12 slices of pizza. Each pizza is cut into 4 slices. How many pizzas are there? Write a number sentence to solve.

22. A van has 4 rows of seats, and each row holds 3 people. How many people can the van hold? Write a number sentence to solve.

Brain Builders

23. Mika rides his bike 3 miles each way to his friend's house. He leaves his house at 4 P.M. If Mika rides his bike to and from his friend's house 2 days a week, how many miles does he ride in a week? What information is extra?

24. Eli and 3 friends divide 24 large marshmallows. Each s'more uses 2 marshmallows. How many s'mores can each camper have?

25. **Test Practice** Mr. Thompson bought 3 of the same item. He paid a total of $21. Which item did he buy?

Ⓐ  $6

Ⓒ  $2

Ⓑ  $7

Ⓓ  $75

Reflect

Chapter 7
Answering the
ESSENTIAL QUESTION

Use what you learned about multiplication and division to complete the graphic organizer.

- **Draw an Array**
- **Repeated Subtraction**
- **Double a Known Fact**
- **Inverse Operations**

ESSENTIAL QUESTION
What strategies can be used to learn multiplication and division facts?

Reflect on the ESSENTIAL QUESTION Write your answer below.

_____

420 Chapter 7 Multiplication and Division

Name _____   Date _____

Score _____

Performance Task

Brain Builders

At the Movies

The movie theater sells bags of popcorn. The prices of the popcorn are listed in the table.

Plain Popcorn	Buttered Popcorn	Caramel Popcorn
$2 per bag	$3 per bag	$4 per bag

Show all your work to receive full credit.

Part A

At the first showing of the movie, the snack stand sold 5 bags of buttered popcorn. How much money did they make? Show your work.

Part B

That same showing, the snack stand sold $32 worth of caramel popcorn. How many bags of caramel popcorn did they sell? Show your work.

Online Content at connectED.mcgraw-hill.com   Performance Task 420PT1

Part C

At the evening showing of the movie, the snack stand sold $12 worth of popcorn. Write three different possibilities for the number of bags of each type of popcorn that could have been sold.

Part D

At the late night showing, the snack stand sold 8 bags of plain popcorn, 6 bags of buttered popcorn, and 7 bags of caramel popcorn. Fill in the table to show how much the snack stand earned from each kind of popcorn. Then tell how much money they earned in all.

	Bags Sold	Money Earned
Plain Popcorn	8	
Buttered Popcorn	6	
Caramel Popcorn	7	

Total Money Earned: _____

Part E

During the weekend afternoon movie, the snack stand sold 9 bags of plain popcorn, 0 bags of buttered popcorn, and 8 bags of caramel popcorn. How much money did the stand earn in all? Show your work.

Glossary/Glosario

Go online for the eGlossary.

Go to the **eGlossary** to find out more about these words in the following 13 languages:

Arabic · Bengali · Brazilian Portuguese · Cantonese · English · Haitian Creole
Hmong · Korean · Russian · Spanish · Tagalog · Urdu · Vietnamese

English	Spanish/Español

analog clock A clock that has an *hour* hand and a *minute* hand.

reloj analógico Reloj que tiene una manecilla *horaria* y un *minutero*.

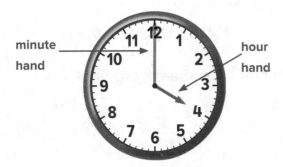 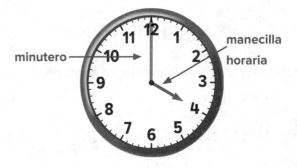

analyze To break information into parts and study it.

analizar Separar la información en partes y estudiarla.

angle A figure that is formed by two *rays* with the same *endpoint*.

ángulo Figura formada por dos *semirrectas* con el mismo *extremo*.

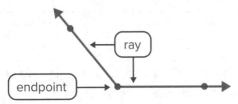 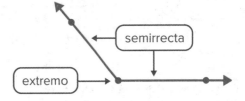

area The number of *square units* needed to cover the inside of a region or *plane figure*.

área Cantidad de *unidades cuadradas* necesarias para cubrir el interior de una región o *figura plana*.

area = 6 square units

área = 6 unidades cuadradas

Online Content at connectED.mcgraw-hill.com

Glossary/Glosario **GL1**

Aa

array Objects or symbols displayed in rows of the same *length* and columns of the same *length*.

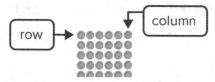

Associative Property of Addition The property that states that the grouping of the addends does not change the sum.

$(4 + 5) + 2 = 4 + (5 + 2)$

Associative Property of Multiplication The property that states that the grouping of the *factors* does not change the *product*.

$3 \times (6 \times 2) = (3 \times 6) \times 2$

attribute A characteristic of a shape.

arreglo Objetos o símbolos organizados en filas y columnas de la misma *longitud*.

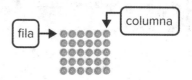

propiedad asociativa de la suma Propiedad que establece que la forma de agrupar los sumandos no altera la suma.

$(4 + 5) + 2 = 4 + (5 + 2)$

propiedad asociativa de la multiplicación Propiedad que establece que la forma de agrupar los *factores* no altera el *producto*.

$3 \times (6 \times 2) = (3 \times 6) \times 2$

atributo Característica de una figura.

Bb

bar diagram A problem-solving strategy in which bar models are used to visually organize the facts in a problem.

|---------- 96 children ----------|
girls	boys
|------ 60 ------|---- ? ----|

bar graph A *graph* that compares *data* by using bars of different *lengths* or heights to show the values.

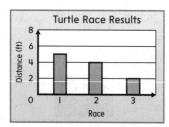

diagrama de barras Estrategia para la resolución de problemas en la cual se usan barras para modelos de organizar visualmente los datos de un problema.

|---------- 96 estudiantes ----------|
niñas	niños
|------ 60 ------|---- ? ----|

gráfica de barras *Gráfica* en la que se comparan *los datos* con barras de distintas *longitudes* o alturas para ilustrar los valores.

Cc

capacity The amount a container can hold, measured in *units* of dry or liquid measure.

combination A new set made by combining parts from other sets.

Commutative Property of Addition The property that states that the order in which two numbers are added does not change the *sum*.

$$12 + 15 = 15 + 12$$

Commutative Property of Multiplication The property that states that the order in which two numbers are multiplied does not change the *product*.

$$7 \times 2 = 2 \times 7$$

compose To form by putting together.

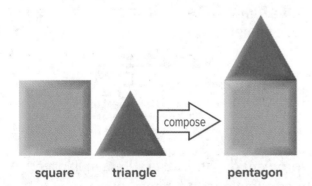

square triangle pentagon

composite figure A figure made up of two or more shapes.

capacidad Cantidad que puede contener un recipiente, medida en *unidades* líquidas o secas.

combinación Conjunto nuevo que se forma al combinar partes de otros conjuntos.

propiedad conmutativa de la suma Propiedad que establece que el orden en el cual se suman dos o más números no altera la *suma*.

$$12 + 15 = 15 + 12$$

propiedad conmutativa de la multiplicación Propiedad que establece que el orden en el cual se multiplican dos o más números no altera el *producto*.

$$7 \times 2 = 2 \times 7$$

componer Juntar para formar.

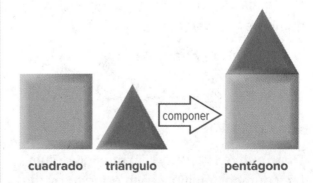

cuadrado triángulo pentágono

figura compuesta Figura conformada por dos o más figuras.

Dd

data Numbers or symbols sometimes collected from a *survey* or experiment to show information. *Datum* is singular; *data* is plural.

decagon A *polygon* with 10 sides and 10 *angles*.

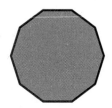

decompose To break a number into different parts.

denominator The bottom number in a *fraction*.

In $\frac{5}{6}$, 6 is the denominator.

digit A symbol used to write a number. The ten digits are 0, 1, 2, 3, 4, 5, 6, 7, 8, and 9.

digital clock A clock that uses only numbers to show time.

Distributive Property To multiply a sum by a number, multiply each *addend* by the number and add the *products*.

$4 \times (1 + 3) = (4 \times 1) + (4 \times 3)$

---

datos Números o símbolos que se recopilan mediante una *encuesta* o un experimento para mostrar información.

decágono *Polígono* con 10 lados y 10 *ángulos*.

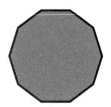

descomponer Separar un número en diferentes partes.

denominador El número de abajo en una *fracción*.

En $\frac{5}{6}$, 6 es el denominador.

dígito Símbolo que se usa para escribir un número. Los diez dígitos son 0, 1, 2, 3, 4, 5, 6, 7, 8 y 9.

reloj digital Reloj que marca la hora solo con números.

propiedad distributiva Para multiplicar una suma por un número, puedes multiplicar cada *sumando* por el número y luego sumar los *productos*.

$4 \times (1 + 3) = (4 \times 1) + (4 \times 3)$

divide (division) To separate into equal groups, to find the number of groups, or the number in each group.

dividend A number that is being *divided*.

$3\overline{)9}$ 9 is the dividend.

division sentence A *number sentence* that uses the *operation* of *division*.

divisor The number by which the *dividend* is being *divided*.

$3\overline{)9}$ 3 is the divisor.

double Twice the number or amount.

dividir (división) Separar en grupos iguales para hallar el número de grupos que hay, o el número de elementos que hay en cada grupo.

dividendo Número que se *divide*.

$3\overline{)9}$ 9 es el dividendo.

división Enunciado *numérico* que usa la *operación* de *dividir*.

divisor Número entre el cual se *divide* el *dividendo*.

$3\overline{)9}$ 3 es el divisor.

doble Dos veces el número o la cantidad.

Ee

elapsed time The amount of time that has passed from the beginning to the end of an activity.

endpoint The point at the beginning of a *ray*.

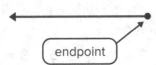

equal groups Groups that have the same number of objects.

equation A *number sentence* that contains an equals sign, =, indicating that the left side of the equals sign has the same value as the right side.

equivalent fractions *Fractions* that have the same value.

$\frac{2}{4} = \frac{1}{2}$

tiempo transcurrido Cantidad de tiempo que ha pasado entre el principio y el fin de una actividad.

extremo Punto al principio de una *semirrecta*.

grupos iguales Grupos que tienen el mismo número de objetos.

ecuación *Enunciado* numérico que tiene un signo igual, =, e indica que el lado izquierdo del signo igual tiene el mismo valor que el lado derecho.

fracciones equivalentes *Fracciones* que tienen el mismo valor.

$\frac{2}{4} = \frac{1}{2}$

Ee

estimate A number close to an exact value. An estimate indicates *about* how much.

$$47 + 22 \text{ is about } 70.$$

evaluate To find the value of an *expression* by replacing *variables* with numbers.

expanded form/expanded notation The representation of a number as a sum that shows the value of each *digit*.

$$536 \text{ is written as } 500 + 30 + 6.$$

experiment To test an idea.

expression A combination of numbers and *operations*.

$$5 + 7$$

estimación Número cercano a un valor exacto. Una estimación indica una cantidad *aproximada*.

$$47 + 22 \text{ es aproximadamente } 70.$$

evaluar Calcular el valor de una *expresión* reemplazando las *variables* por números.

forma desarrollada/notación desarrollada Representación de un número como la suma que muestra el valor de cada *dígito*.

$$536 \text{ se escribe como } 500 + 30 + 6.$$

experimentar Probar una idea.

expresión Combinación de números y *operaciones*.

$$5 + 7$$

Ff

fact family A group of *related facts* using the same numbers.

$5 + 3 = 8$	$5 \times 3 = 15$
$3 + 5 = 8$	$3 \times 5 = 15$
$8 - 3 = 5$	$15 \div 5 = 3$
$8 - 5 = 3$	$15 \div 3 = 5$

factor A number that is *multiplied* by another number.

foot (ft) A customary unit for measuring *length*. Plural is *feet*.

$$1 \text{ foot} = 12 \text{ inches}$$

formula An *equation* that shows the relationship between two or more quantities.

familia de operaciones Grupo de *operaciones relacionadas* que tienen los mismos números.

$5 + 3 = 8$	$5 \times 3 = 15$
$3 + 5 = 8$	$3 \times 5 = 15$
$8 - 3 = 5$	$15 \div 5 = 3$
$8 - 5 = 3$	$15 \div 3 = 5$

factor Número que se *multiplica* por otro número.

pie Unidad usual para medir la *longitud*.

$$1 \text{ pie} = 12 \text{ pulgadas}$$

fórmula *Ecuación* que muestra la relación entre dos o más cantidades.

fraction A number that represents part of a whole or part of a set.

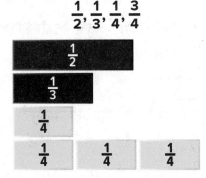

frequency table A *table* for organizing a set of *data* that shows the number of times each result has occurred.

Bought Lunch Last Month	
Name	Frequency
Julia	6
Martin	4
Lin	5
Tanya	4

fracción Número que representa una parte de un todo o una parte de un conjunto.

$$\frac{1}{2}, \frac{1}{3}, \frac{1}{4}, \frac{3}{4}$$

tabla de frecuencias *Tabla* para organizar un conjunto de *datos* que muestra el número de veces que ha ocurrido cada resultado.

Compraron almuerzo el mes pasado	
Nombre	Frecuencia
Julia	6
Martín	4
Lin	5
Tanya	4

gram (g) A *metric unit* for measuring lesser *mass*.

graph An organized drawing that shows sets of *data* and how they are related to each other. Also a type of chart.

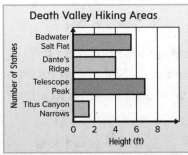

bar graph

gramo (g) *Unidad métrica* para medir la *masa*.

gráfica Dibujo organizado que muestra conjuntos de *datos* y cómo se relacionan. También, es un tipo de diagrama.

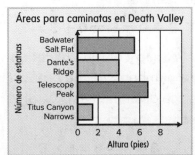

gráfica de barras

Hh

half inch $(\frac{1}{2})$ One of two equal parts of an *inch*.

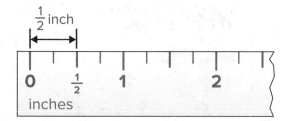

hexagon A *polygon* with six *sides* and six *angles*.

hour (h) A *unit* of time equal to 60 *minutes*.

1 hour = 60 minutes

hundreds A position of *place value* that represents the numbers 100–999.

media pulgada $(\frac{1}{2})$ Una de dos partes iguales de una *pulgada*.

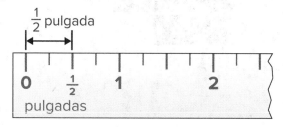

hexágono *Polígono* con seis *lados* y seis *ángulos*.

hora (h) *Unidad* de tiempo igual a 60 *minutos*.

1 hora = 60 minutos

centenas *Valor posicional* que representa los números del 100 al 999.

Ii

Identity Property of Addition If you add zero to a number, the sum is the same as the given number.

$$3 + 0 = 3 \text{ or } 0 + 3 = 3$$

Identity Property of Multiplication If you *multiply* a number by 1, the *product* is the same as the given number.

$$8 \times 1 = 8 = 1 \times 8$$

propiedad de identidad de la suma Si sumas cero a un número, la suma es igual al número dado.

$$3 + 0 = 3 \text{ o } 0 + 3 = 3$$

propiedad de identidad de la multiplicación Si *multiplicas* un número por 1, el *producto* es igual al número dado.

$$8 \times 1 = 8 = 1 \times 8$$

interpret To take meaning from information.

inverse operations *Operations* that undo each other.

Addition and subtraction are inverse, or opposite, operations.

Multiplication and *division* are also inverse operations.

is equal to (=) Having the same value.

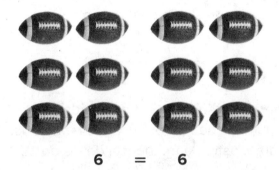

6 is equal to, or the same, as 6.

is greater than > An inequality relationship showing that the value on the left of the symbol is greater than the value on the right.

5 > 3 5 is greater than 3.

is less than < An inequality relationship showing that the value on the left side of the symbol is smaller than the value on the right side.

4 < 7 4 is less than 7.

interpretar Extraer significado de la información.

operaciones inversas *Operaciones* que se anulan entre sí.

La suma y la resta son operaciones inversas u opuestas.

La *multiplicación* y la *división* también son operaciones inversas.

es igual a (=) Que tienen el mismo valor.

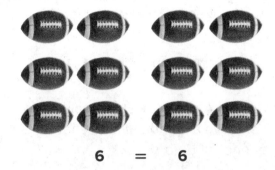

6 es igual o lo mismo que 6.

es mayor que > Relación de desigualdad que muestra que el valor a la izquierda del signo es más grande que el valor a la derecha.

5 > 3 5 es mayor que 3.

es menor que < Relación de desigualdad que muestra que el valor a la izquierda del signo es más pequeño que el valor a la derecha.

4 < 7 4 es menor que 7.

key Tells what or how many each symbol in a *graph* stands for.

kilogram (kg) A *metric unit* for measuring greater *mass*.

clave Indica qué significa o cuánto representa cada símbolo en una *gráfica*.

kilogramo (kg) *Unidad métrica* para medir la *masa*.

known fact A fact that you already know.

hecho conocido Hecho que ya sabes.

length Measurement of the distance between two *points*.

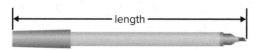

line A straight set of *points* that extend in opposite directions without ending.

line plot A graph that uses columns of Xs above a *number line* to show frequency of *data*.

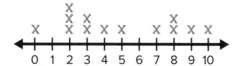

liquid volume The amount of liquid a container can hold. Also known as *capacity*.

liter (L) A *metric unit* for measuring greater *volume* or *capacity*.

1 liter = 1,000 milliliters

longitud Medida de la distancia entre dos *puntos*.

recta Conjunto de *puntos* alineados que se extiende sin fin en direcciones opuestas.

diagrama lineal Gráfica que usa columnas de X sobre una *recta numérica* para mostrar la frecuencia de los *datos*.

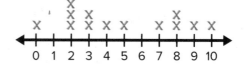

volumen líquido Cantidad de líquido que puede contener un recipiente. También se conoce como *capacidad*.

litro (L) *Unidad métrica* para medir el *volumen* o la *capacidad*.

1 litro = 1,000 mililitros

Mm

mass The amount of matter in an object. Two examples of *units* of mass are *gram* and *kilogram*.

masa Cantidad de materia en un cuerpo. Dos ejemplos de *unidades* de masa son el *gramo* y el *kilogramo*.

mental math Ordering or grouping numbers so that they are easier to compute in your head.

cálculo mental Ordenar o agrupar números de modo que sean más fáciles de operar mentalmente.

Mm

metric system (SI) The measurement system based on powers of 10 that includes *units* such as *meter, gram,* and *liter.*

metric unit A *unit* of measure in the *metric system*.

milliliter (mL) A *metric unit* used for measuring lesser *capacity*.

1,000 milliliters = 1 liter

minute (min) A *unit* used to measure short periods of time.

1 minute = 60 seconds

multiple A multiple of a number is the *product* of that number and any *whole number*.

15 is a multiple of 5 because 3 × 5 = 15.

multiplication An *operation* on two numbers to find their *product*. It can be thought of as repeated addition.

$3 \times 4 = 12$

$4 + 4 + 4 = 12$

multiplication sentence A *number sentence* that uses the *operation* of *multiplication*.

multiply To find the *product* of 2 or more numbers.

---

sistema métrico (SI) Sistema decimal de medidas que se basa en potencias de 10 y que incluye *unidades* como el *metro,* el *gramo* y el *litro*.

unidad métrica *Unidad* de medida del *sistema métrico*.

mililitro (mL) *Unidad métrica* para medir las *capacidades* pequeñas.

1,000 mililitros = 1 litro

minuto (min) *Unidad* que se usa para medir el tiempo.

1 minuto = 60 segundos

múltiplo Un múltplo de un número es el *producto* de ese número y cualquier otro *número natural*.

15 es múltiplo de 5 porque 3 × 5 = 15.

multiplicación *Operación* entre dos números para hallar su *producto*. Puede considerar como una suma repetida.

$3 \times 4 = 12$

$4 + 4 + 4 = 12$

multiplicación *Enunciado numérico* que usa la *operación* de *multiplicar*.

multiplicar Hallar el *producto* de 2 o más números.

Nn

number line A line with numbers marked in order and at regular intervals.

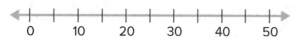

recta numérica Recta con números marca dos en orden y a intervalos regulares.

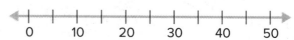

number sentence An *expression* using numbers and the =, <, or > sign.

$$5 + 4 = 9; 8 > 5$$

enunciado numérico *Expresión* que usa números y el signo =, <, o >.

$$5 + 4 = 9; 8 > 5$$

numerator The number above the bar in a *fraction*; the part of the *fraction* that tells how many of the equal parts are being used.

In the fraction $\frac{3}{4}$, 3 is the numerator.

numerador Número que está encima de la barra de *fracción*; la parte de la *fracción* que indica cuántas partes iguales se están usando.

En la fracción $\frac{3}{4}$, 3 es numerador.

Oo

observe A method of collecting *data* by watching.

observar Método que utiliza la observación para recopilar *datos*.

octagon A *polygon* with eight sides and eight *angles*.

octágono *Polígono* de ocho lados y ocho ángulos.

operation(s) A mathematical process such as addition (+), subtraction (−), multiplication (×), and division (÷).

operación Proceso matemático como la suma (+), la resta (−), la *multiplicación* (×) y la *división* (÷).

Pp

parallel (lines) *Lines* that are the same distance apart. Parallel lines do not meet.

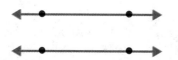

rectas paralelas *Rectas* separadas por la misma distancia en cualquier punto. Las rectas paralelas no se intersecan.

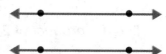

parallelogram A *quadrilateral* with four sides in which each pair of opposite sides is *parallel* and equal in *length*.

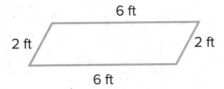

parentheses Symbols that are used to group numbers. They show which *operations* to complete first in a *number sentence*.

partition To *divide* or "break up."

pattern A sequence of numbers, figures, or symbols that follow a rule or design.

2, 4, 6, 8, 10

pentagon A *polygon* with five sides and five *angles*.

perimeter The distance around a shape or region.

period The name given to each group of three *digits* on a *place-value* chart.

pictograph A *graph* that compares *data* by using pictures or symbols.

paralelogramo *Cuadrilátero* en el que cada par de lados opuestos son *paralelos* y tienen la misma *longitud*.

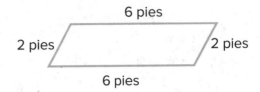

paréntesis Signos que se usan para agrupar números. Muestran cuáles *operaciones* se completan primero en un *enunciado numérico*.

separar *Dividir* o desunir.

patrón Sucesión de números, figuras o símbolos que sigue una regla o un diseño.

2, 4, 6, 8, 10

pentágono *Polígono* de cinco lados y cinco ángulos.

perímetro Distancia alrededor de una figura o región.

período Nombre dado a cada grupo de tres *dígitos* en una tabla de valor *posicional*..

pictografía *Gráfica* en la que se comparan *datos* usando figuras o símbolos.

Pp

picture graph A *graph* that has different pictures to show information collected.

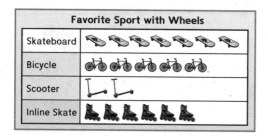

gráfica con imágenes *Gráfica* que tiene diferentes imágenes para ilustrar la información recopilada.

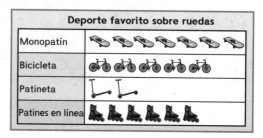

place value The value given to a *digit* by its place in a number.

valor posicional Valor dado a un *dígito* según su lugar en el número.

plane figure A *two-dimensional figure* that lies entirely within one plane, such as a *triangle* or *square*.

figura plana *Figura bidimensional* que yace completamente en un plano, como un *triángulo* o un *cuadrado*.

point An exact location in space.

punto Ubicación exacta en el espacio.

polygon A closed *plane figure* formed by *line* segments that meet only at their *endpoints*.

polígono *Figura plana* cerrada formada por segmentos de *recta* que solo se unen en sus *extremos*.

prediction Something you think will happen, such as a specific outcome of an *experiment*.

predicción Algo que crees que sucederá, como un resultado específico de un *experimento*.

product The answer to a *multiplication* problem.

producto Respuesta a un problema de *multiplicación*.

Qq

quadrilateral A shape that has 4 sides and 4 *angles*.

square   rectangle   parallelogram

cuadrilátero Figura que tiene 4 lados y 4 *ángulos*.

cuadrado   rectángulo   paralelogramo

GL14 Glossary/Glosario

Qq

quarter hour One-fourth of an *hour*, or 15 *minutes*.

cuarto de hora La cuarta parte de una *hora* o 15 *minutos*.

quarter inch $\left(\frac{1}{4}\right)$ One of four equal parts of an *inch*.

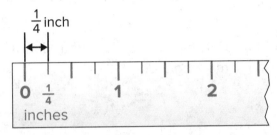

cuarto de pulgada $\left(\frac{1}{4}\right)$ Una de cuatro partes iguales de una *pulgada*.

quotient The answer to a *division* problem.

$15 \div 3 = 5$ ← 5 is the quotient.

cociente Respuesta a un problema de *división*.

$15 \div 3 = 5$ ← 5 es el cociente.

Rr

ray A part of a *line* that has one *endpoint* and extends in one direction without ending.

semirrecta Parte de una *recta* que tiene un *extremo* y que se extiende sin fin en una dirección.

reasonable Within the bounds of making sense.

razonable Dentro de los límites de lo que tiene sentido.

rectangle A *quadrilateral* with four *right angles*; opposite sides are equal in *length* and are *parallel*.

rectangle *Cuadrilátero* con cuatro *ángulos rectos*; los lados opuestos son de igual *longitud* y *paralelos*.

regroup To use *place value* to exchange equal amounts when renaming a number.

reagrupar Usar el *valor posicional* para intercambiar cantidades iguales cuando se convierte un número.

Online Content at connectED.mcgraw-hill.com

Glossary/Glosario **GL15**

Rr

related fact(s) Basic facts using the same numbers. Sometimes called a *fact family*.

$4 + 1 = 5$	$5 \times 6 = 30$
$1 + 4 = 5$	$6 \times 5 = 30$
$5 - 4 = 1$	$30 \div 5 = 6$
$5 - 1 = 4$	$30 \div 6 = 5$

relacionadas Operaciones básicas que tienen los mismos números. También se llaman *familia de operaciones*.

$4 + 1 = 5$	$5 \times 6 = 30$
$1 + 4 = 5$	$6 \times 5 = 30$
$5 - 4 = 1$	$30 \div 5 = 6$
$5 - 1 = 4$	$30 \div 6 = 5$

repeated subtraction To subtract the same number over and over until you reach 0.

resta repetida Procedimiento por el que se resta un número una y otra vez hasta llegar a 0.

rhombus A *parallelogram* with four sides of the same *length*.

rombo *Paralelogramo* con cuatro lados de la misma *longitud*.

right angle An *angle* that forms a *square* corner.

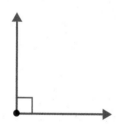

ángulo recto *Ángulo* que forma una esquina *cuadrada*.

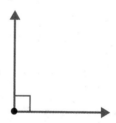

right triangle A *triangle* with one *right angle*.

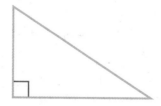

triángulo rectángulo *Triángulo* con un *ángulo recto*.

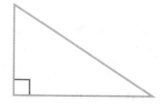

round To change the value of a number to one that is easier to work with. To find the nearest value of a number based on a given *place value*. 27 rounded to the nearest ten is 30.

redondear Cambiar el valor de un número a uno con el que es más fácil trabajar. Hallar el valor más cercano a un número con base en un *valor posicional* dado. 27 redondeado a la décima más cercana es 30.

Ss

scale A set of numbers that represents the *data* in a *graph*.

square A *plane* shape that has four equal sides. Also a *rectangle*.

square unit A *unit* for measuring *area*.

standard form/standard notation The usual way of writing a number that shows only its *digits*, no words.

537   89   1642

survey A method of collecting *data* by asking a group of people a question.

escala Conjunto de números que representa los *datos* en una *gráfica*.

cuadrado *Figura plana* que tiene cuatro lados iguales. También es un *rectángulo*.

unidad cuadrada *Unidad* para medir el *área*.

forma estándar/notación estándar Manera habitual de escribir un número usando solor sus *dígitos*, sin usar palabras.

537   89   1642

encuesta Método para recopilar *datos* haciendo una pregunta a un grupo de personas.

Tt

table A way to organize and display *data* in rows and columns.

tally chart A way to keep track of *data* using *tally marks* to record the results.

| What is Your Favorite Color? ||
Color	Tally
Blue	ͲͲ lll
Green	llll

tabla Manera de organizar y representar *datos* en filas y columnas.

tabla de conteo Manera de llevar la cuenta de los *datos* usando *marcas de conteo* para anotar los resultados.

| ¿Cuál es tu color favorito? ||
Color	Conteo
azul	ͲͲ lll
verde	llll

Glossary/Glosario **GL17**

Tt

tally mark(s) A mark made to record and display *data* from a *survey*.

marca de conteo Marca que se hace para anotar y presentar los *datos* de una.

thousands A position of *place value* that represents the numbers 1,000–9,999.

In 1,253, the 1 is in the thousands place.

millares *Valor posicional* que representa los números del 1,000 al 9,999.

En 1,253, el 1 está en el lugar de los millares.

time interval The time that passes from the start of an activity to the end of an activity.

intervalo de tiempo Tiempo que transcurre entre de el comienzo y el final de una actividad.

time line A *number line* that shows when and in what order events took place.

Jason's Time Line

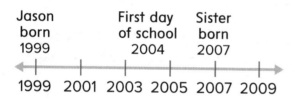

línea cronológica *Recta numérica* que muestra cuándo y en qué orden ocurrieron los eventos.

Linea Cronológica de Jason

trapezoid A *quadrilateral* with exactly one pair of *parallel* sides.

trapecio *Cuadrilátero* con exactamente un par de lados *paralelos*.

tree diagram A branching diagram that shows all the possible *combinations* when combining sets.

diagrama de árbol Diagrama con ramas que muestra todas las posibles *combinaciones* al reunir conjuntos.

triangle A *polygon* with three sides and three *angles*.

triángulo *Polígono* con tres lados y tres *ángulos*.

two-dimensional figure The outline of a shape—such as a *triangle*, *square*, or *rectangle*—that has only *length*, *width*, and *area*. Also called a *plane figure*.

figura bidimensional Contorno de una figura, como un *triángulo*, un *cuadrado* o un *rectángulo*, que solo tiene *largo*, *ancho* y *área*. También conocida como *figura plana*.

Uu

unit The quantity of 1, usually used in reference to measurement.

unidad Cantidad unitaria, que se usa para referirse a medidas.

unit fraction Any *fraction* with a *numerator* of 1.

$$\frac{1}{2}, \frac{1}{3}, \frac{1}{4}$$

fracción unitaria Cualquier *fracción* cuyo *numerador* es 1.

$$\frac{1}{2}, \frac{1}{3}, \frac{1}{4}$$

unit square A *square* with a side *length* of one *unit*.

cuadrado unitario *Cuadrado* cuya *longitud* de los lados es igual a una *unidad*.

unknown A missing number, or the number to be solved for.

incógnita Número que falta, o el número por el que hay que resolver algo.

Vv

variable A letter or symbol used to represent an *unknown* quantity.

variable Letra o símbolo que se usa para representar una cantidad *desconocida*.

vertex The *point* where two *rays* meet in an *angle*.

vértice *Punto* donde se unen dos semirrectas y forman un *ángulo*.

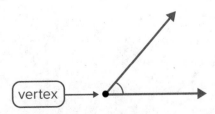

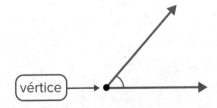

Ww

whole number The numbers 0, 1, 2, 3, 4 . . .

word form/word notation The form of a number that uses written words.

6,472

six thousand, four hundred seventy-two

número natural Los números 0, 1, 2, 3, 4 . . .

forma verbal/notación verbal Forma de un número que se escribe en palabras.

6,472

seis mil cuatrocientos setenta y dos

Yy

yard (yd) A customary unit for measuring *length*.

1 yard = 3 feet or 36 inches

yarda (yd) Unidad usual para medir la *longitud*.

1 yarda = 3 pies o 36 pulgadas

Zz

Zero Property of Multiplication The property that states that any number multiplied by zero is zero.

$0 \times 5 = 0 \quad 5 \times 0 = 0$

propiedad del cero de la multiplicación Propiedad que establece que cualquier número multiplicado por cero es igual a cero.

$0 \times 5 = 0 \quad 5 \times 0 = 0$

Name ....

Work Mat 1: Thousands Place-Value Chart

thousands	hundreds	tens	ones

Work Mat 1 Thousands Place-Value Chart **WM1**

Work Mat 2: Number Lines

0 1 2 3 4 5 6 7 8 9 10

WM2 **Work Mat 2** Number Lines

Work Mat 3: Hundred Chart

1	2	3	4	5	6	7	8	9	10
11	12	13	14	15	16	17	18	19	20
21	22	23	24	25	26	27	28	29	30
31	32	33	34	35	36	37	38	39	40
41	42	43	44	45	46	47	48	49	50
51	52	53	54	55	56	57	58	59	60
61	62	63	64	65	66	67	68	69	70
71	72	73	74	75	76	77	78	79	80
81	82	83	84	85	86	87	88	89	90
91	92	93	94	95	96	97	98	99	100

Work Mat 4: Place-Value Chart

thousands	hundreds	tens	ones

WM4 **Work Mat 4** Place-Value Chart

Name

Work Mat 5: Centimeter Grid

Work Mat 5 Centimeter Grid **WM5**

Work Mat 6: Bar Diagram

WM6   Work Mat 6 Bar Diagram

Name _____

Work Mat 7: Multiplication Fact Table, to 12

×	0	1	2	3	4	5	6	7	8	9	10	11	12
0	0	0	0	0	0	0	0	0	0	0	0	0	0
1	0	1	2	3	4	5	6	7	8	9	10	11	12
2	0	2	4	6	8	10	12	14	16	18	20	22	24
3	0	3	6	9	12	15	18	21	24	27	30	33	36
4	0	4	8	12	16	20	24	28	32	36	40	44	48
5	0	5	10	15	20	25	30	35	40	45	50	55	60
6	0	6	12	18	24	30	36	42	48	54	60	66	72
7	0	7	14	21	28	35	42	49	56	63	70	77	84
8	0	8	16	24	32	40	48	56	64	72	80	88	96
9	0	9	18	27	36	45	54	63	72	81	90	99	108
10	0	10	20	30	40	50	60	70	80	90	100	110	120
11	0	11	22	33	44	55	66	77	88	99	110	121	132
12	0	12	24	36	48	60	72	84	96	108	120	132	144

Work Mat 8: Algebra Mat

=

WM8 Work Mat 8 Algebra Mat